Chemistry Research and Applications

Chemistry Research and Applications

Pyrimidines and their Importance
Roger G. Ward (Editor)
2023. ISBN: 979-8-88697-656-4 (Softcover)
2023. ISBN: 979-8-88697-663-2 (eBook)

What to Know about Lanthanum
Catherine C. Bradley (Editor)
2023. ISBN: 979-8-88697-615-1 (Softcover)
2023. ISBN: 979-8-88697-623-6 (eBook)

The Future of Biorefineries
Waldemar Nyström (Editor)
2023. ISBN: 979-8-88697-524-6 (Hardcover)
2023. ISBN: 979-8-88697-528-4 (eBook)

Properties and Uses of Antimony
David J. Jenkins (Editor)
2022. ISBN: 979-8-88697-081-4 (Softcover)
2022. ISBN: 979-8-88697-088-3 (eBook)

The Science of Carbamates
Güllü Kaymak (Editor)
2022. ISBN: 978-1-68507-708-2 (Softcover)
2022. ISBN: 978-1-68507-872-0 (eBook)

Deep Eutectic Solvents: Properties, Applications and Toxicity
Carlos Eduardo de Araújo Padilha, PhD, Everaldo Silvino dos Santos, PhD, Francisco Canindé de Sousa Júnior, PhD, Nathália Saraiva Rios, PhD (Editors)
2022. ISBN: 978-1-68507-719-8 (Hardcover)
2022. ISBN: 978-1-68507-799-0 (eBook)

More information about this series can be found at
https://novapublishers.com/product-category/series/chemistry-research-and-applications/

Wilbur M. Hulett
Editor

Alkali Metals

New Research

Copyright © 2023 by Nova Science Publishers, Inc.

All rights reserved. No part of this book may be reproduced, stored in a retrieval system or transmitted in any form or by any means: electronic, electrostatic, magnetic, tape, mechanical photocopying, recording or otherwise without the written permission of the Publisher.

We have partnered with Copyright Clearance Center to make it easy for you to obtain permissions to reuse content from this publication. Please visit copyright.com and search by Title, ISBN, or ISSN.

For further questions about using the service on copyright.com, please contact:

Copyright Clearance Center
Phone: +1-(978) 750-8400 Fax: +1-(978) 750-4470 E-mail: info@copyright.com

NOTICE TO THE READER

The Publisher has taken reasonable care in the preparation of this book but makes no expressed or implied warranty of any kind and assumes no responsibility for any errors or omissions. No liability is assumed for incidental or consequential damages in connection with or arising out of information contained in this book. The Publisher shall not be liable for any special, consequential, or exemplary damages resulting, in whole or in part, from the readers' use of, or reliance upon, this material. Any parts of this book based on government reports are so indicated and copyright is claimed for those parts to the extent applicable to compilations of such works.

Independent verification should be sought for any data, advice or recommendations contained in this book. In addition, no responsibility is assumed by the Publisher for any injury and/or damage to persons or property arising from any methods, products, instructions, ideas or otherwise contained in this publication.

This publication is designed to provide accurate and authoritative information with regards to the subject matter covered herein. It is sold with the clear understanding that the Publisher is not engaged in rendering legal or any other professional services. If legal or any other expert assistance is required, the services of a competent person should be sought. FROM A DECLARATION OF PARTICIPANTS JOINTLY ADOPTED BY A COMMITTEE OF THE AMERICAN BAR ASSOCIATION AND A COMMITTEE OF PUBLISHERS.

Library of Congress Cataloging-in-Publication Data

ISBN: 979-8-88697-706-6

Published by Nova Science Publishers, Inc. † New York

Contents

Preface vii

Chapter 1   **The Interaction between Spin Polarized Alkali Atoms: Shifts of the Magnetic Resonance Lines** 1
Victor Kartoshkin

Chapter 2   **Effects of Alkali Elements on Perovskite Halide Compounds** 33
Takeo Oku, Naoki Ueoka, Hayato Machiba
Atsushi Suzuki, Riku Okumura and Ayu Enomoto

Chapter 3   **The Interaction between Spin Polarized Cesium Atoms and Alkali Atoms: Spin Exchange Collisions** 59
Victor A. Kartoshkin

Chapter 4   **Superconducting State Parameters of Alkali Metals** 83
Rajesh C. Malan and Aditya M. Vora

Index 97

Preface

This book contains four chapters on new research regarding alkali metals. Chapter One examines the interaction between spin polarized alkali atoms, specifically the shifts of the magnetic resonance lines. Chapter Two reviews the effects of alkali elements on perovskite halide compounds. Chapter Three examines the interaction between spin polarized cesium atoms and alkali atoms, specifically spin exchange collisions. Chapter Four looks at the superconducting state parameters of alkali metals.

Chapter 1 - At present, devices based on the principle of optical orientation of atoms play a significant role in physical investigations. In various magnetic measurements, quantum magnetometers with optical pumping are used; optical orientation of atoms is employed in the development of quantum gyroscopes, as well as in magnetoencephalographs. The role of working media in such devices is played, in particular, by alkali metal atoms in the ground state.

Atoms can be of one species as well as mixtures of different alkali atoms. Alkali atoms in the absorption chamber of such devices collide with one another so that collisions of alkali atoms are accompanied with the well-known process of spin exchange (i.e., exchange of the electron polarization between colliding atoms). The spin exchange process substantially affects the magnetic resonance linewidth, as well as the magnetic resonance frequency shift of alkali atoms, participating in a collision. If one of the colliding atoms has been preliminarily polarized, its polarization can be transferred to the partner of the collision. In particular, this makes it possible to polarize the collision partner "indirectly" without using a polarized resonance optical radiation. Such an indirect polarization is used, among other things, in the situation in which "direct" optical orientation cannot be realized for some reason or when it is necessary to isolate the pumping channels from the magnetic resonance detection channels. The spin exchange process is known to be accompanied by the aforementioned polarization transfer, as well as by the magnetic resonance frequency shift of atoms. When the spin exchange occurs in a mixture of alkali atoms, both collisions between identical atoms and the collisions with atoms of another species affect the value of the frequency shift. Spin-exchange magnetic resonance frequency shifts in a mixture of alkali atoms in the buffer gas atmosphere were considered in.

Chapter 2 - Effects of doping with alkali elements such as Cs, Rb, K, or Na cations on CH3NH3PbI3 perovskite photovoltaic cells were investigated and described. Lattice constants were slightly decreased and increased by K and Na doping, respectively, which indicated that Na atoms occupied interstitial sites in the perovskite crystal. Perovskite solutions with K and formamidinium (FA) iodides added were also used to fabricate perovskite solar cells. The addition of the K salts resulted in fine grained FA added perovskite crystals that were separated by very small gaps. These materials were used to fabricate solar cells that improved short circuit current densities, which resulted in an enhanced conversion efficiency, compared with standard materials. Effects of addition of alkali metal elements to Cu-doped CH3NH3PbI3 photovoltaic devices were also investigated. The series resistance was decreased by simultaneous addition of CuBr2 and RbI, which increased the external quantum efficiencies in the range of 300–500 nm, and the short-circuit current density. The stabilities can also be estimated by first-principle calculations, and the Gibbs energies were decreased by incorporation of alkali metal elements into the perovskite crystals. The copper d-orbital band was slightly above the valence-band maximum and functioned as an acceptor level for carrier generation. Excitation from iodine p-orbitals and copper d-orbitals to alkali metal s-orbitals could suppress carrier recombination and promote carrier transport.

Chapter 3 - The study of interactions involving spin-polarized atoms began quite a long time ago, but in recent years, interest in these interactions has again become topical. This is due to both the development of experimental technique, in particular, semiconductor lasers, including vertical cavity surface-emitting lasers (VCSEL), which, with their miniature size, allow to achieve a high degree of polarization of alkali atoms, in particular, cesium. This kind of miniaturization made it possible to consider the possibilities of designing devices, the principle of operation of which is based on the optical orientation of atoms. Such devices include quantum gyroscopes, magnetoencephalographs, and quantum magnetometers.

The method of optical orientation of atoms makes it possible to obtain spin-polarized optical particles. The presence of polarized atoms in the working chambers of the above devices leads to the fact that collisions between atoms of the working mixtures make it possible to obtain spin polarized atoms that did not interact with the pumping light (indirect optical orientation of atoms). These collisions lead not only to the transfer of spin polarization to the collision partners, but also to the broadening of the magnetic resonance lines and to a shift of the frequency of the magnetic resonance lines. The paper considers the interaction of spin polarized cesium atoms with alkali atoms Cs, Rb, K, and Li in the ground state.

Collisions of alkali-metal atoms in the ground state with the electron spin $S = 1/2$ are accompanied by exchange of electron coordinates between the colliding particles, which leads to the polarization transfer between them (i.e., to the well-known phenomenon of spin exchange). In addition, along with the polarization transfer from one partner to another, the magnetic resonance lines of colliding atoms are broadened and shifted in spin-exchange collisions. The last two processes depend, in particular, on the complex spin-exchange cross sections. The real part of the cross section determines the so-called "spin exchange cross section," which is responsible for broadening of magnetic resonance lines, while the imaginary part - the shift cross section - governs the frequency shift of magnetic resonance. The spin exchange broadening of a magnetic resonance line affects the precision of such quantum electronics devices as quantum frequency standards and magnetometers.

To describe the spin exchange process, we have to know complex spin exchange cross sections. Complex spin-exchange cross sections are calculated based on the data on the singlet $(X1\Sigma+)$ and triplet $(a3\Sigma+)$ potentials describing the interaction Cs alkali-metal atoms in the ground state with other alkaline atoms. The cross-section values allow to calculate the processes of polarization transfer and the relaxation times, as well as the magnetic resonance frequency shifts caused by the Cs - alkaline atoms spin exchange collisions. The paper presents the interaction potentials of the studied pairs of alkaline atoms and calculates the complex spin exchange cross sections for them. Collisions of alkali-metal atoms in the ground state are considered in the energy interval of 10–4–10–2 a.u.

Chapter 4 - The current article is presented with an aim to investigate screening dependency of the superconducting behaviour of alkali metals. The screening dependent superconducting state parameters for five alkali metals (Li, Na, K, Rb, and Cs) are computed in the present work. The pseudopotential given by Fiolhais and his co-worker along with its universal parameters is used for the entire calculation of superconducting state parameters. Critical/transition temperature (Tc), effective interaction strength (N0V), electron-phonon coupling strength (λ) and Coulomb pseudopotential (μ) are the superconducting state parameters included in the present study. Very weak superconducting behaviour is observed for alkali metals. The screening and exchange and correlation effect dependency of the present results is also discussed in the present article.

Chapter 1

The Interaction between Spin Polarized Alkali Atoms: Shifts of the Magnetic Resonance Lines

Victor Kartoshkin[*]
Division of Plasma Physics, Atomic Physics and Astrophysics, Ioffe Institute,
St. Petersburg, Russia

Abstract

At present, devices based on the principle of optical orientation of atoms play a significant role in physical investigations. In various magnetic measurements, quantum magnetometers with optical pumping are used; optical orientation of atoms is employed in the development of quantum gyroscopes, as well as in magnetoencephalographs. The role of working media in such devices is played, in particular, by alkali metal atoms in the ground state.

Atoms can be of one species as well as mixtures of different alkali atoms. Alkali atoms in the absorption chamber of such devices collide with one another so that collisions of alkali atoms are accompanied with the well-known process of spin exchange (i.e., exchange of the electron polarization between colliding atoms). The spin exchange process substantially affects the magnetic resonance linewidth, as well as the magnetic resonance frequency shift of alkali atoms, participating in a collision. If one of the colliding atoms has been preliminarily polarized, its polarization can be transferred to the partner of the collision. In particular, this makes it possible to polarize the collision partner "indirectly" without using a polarized resonance optical radiation. Such an indirect polarization is used, among other things, in the situation in which "direct" optical orientation cannot be realized for some reason or

---

[*] Corresponding Author's Email: victor.kart@mail.ioffe.ru.

In: Alkali Metals
Editor: Wilbur M. Hulett
ISBN: 979-8-88697-706-6
© 2023 Nova Science Publishers, Inc.

when it is necessary to isolate the pumping channels from the magnetic resonance detection channels. The spin exchange process is known to be accompanied by the aforementioned polarization transfer, as well as by the magnetic resonance frequency shift of atoms. When the spin exchange occurs in a mixture of alkali atoms, both collisions between identical atoms and the collisions with atoms of another species affect the value of the frequency shift. Spin-exchange magnetic resonance frequency shifts in a mixture of alkali atoms in the buffer gas atmosphere were considered in.

Keywords: spin exchange, magnetic resonance, frequency shifts

Introduction

As is well known, spin exchange collisions between alkali atoms play an important role in quantum electronic devices, such as quantum magnetometers and time and frequency standards. Along with the polarization transfer between colliding particles, spin exchange collisions lead to a shift in the magnetic resonance frequency of colliding atoms. The problems of the influence of the spin exchange on the polarization transfer processes and related problems of relaxation have now been studied quite well both experimentally (Happer, 1972; Vanier and Audoin, 1989) and theoretically (including spin exchange processes with the participation of nuclear paramagnetic) (Appelt, et al., 1998; Okunevich, 1987). At the same time, magnetic resonance frequency shifts related to this process have been studied much more poorly. In (Dmitriev and Dovator, 1997; Dmitriev and Dovator, 2007), magnetic resonance frequency shifts were measured for the first time for Cs atoms at spin exchange collisions with Rb atoms (Dmitriev and Dovator, 1997) and for Rb atoms at collisions with Cs atoms (Dmitriev and Dovator, 2007). In (Micalizio et al., 2006), an attempt to calculate cross sections of magnetic resonance frequency shifts for homonuclear pairs of alkali atoms was undertaken. For the calculation and the following estimation of the accuracy of quantum frequency standards estimating equations for the interaction potentials were used. The splitting between the singlet and triplet interaction potentials was estimated based on the ionization potentials of colliding atoms. At the same time, the interaction potentials for the triplet and singlet states of the systems Rb–Rb, Cs–Cs, and Cs–Rb were obtained in (Kartoshkin, 1995). These interaction potentials allow the calculation of frequency shifts interesting for us.

To calculate magnetic resonance frequency shifts and to compare results of calculations with experimental data it is necessary to take into account some peculiarities. In particular, the process of the spin exchange upon collisions of alkali atoms and the influence on the magnetic resonance frequency shifts were considered theoretically in (Okunevich, 1995). It should be noted that the derived equations relating the magnetic resonance frequency shifts caused by spin exchange collisions are obtained with allowance for not only the spin exchange process, but also for the diffusion of atoms to walls of the absorption chamber and collisions with atomic particles of the buffer gas. In addition, these equations take into account the complex system of atomic levels of alkali atoms. In this paper, magnetic resonance frequency shift of Rb (5s $^2S_{1/2}$) atoms in a Cs–Rb mixture of alkali atoms in the atmosphere of the buffer gas N_2 were calculated for the first time.

Magnetic Resonance Frequency Shifts Caused by Spin Exchange

As two atoms of alkali metals being in the ground state collide, the well-known process of spin exchange, that is, the exchange of valence electrons between colliding atoms, occurs. These collisions lead to the polarization transfer between colliding atomic particles. Therefore, they can be used for the transfer of polarization from one colliding atom to another. This indirect polarization of atomic particles is well known and is widely used in the optical orientation of atoms. The spin exchange process can be described by the complex cross section:

$$q^{AB} = \bar{q}^{AB} + i \cdot \bar{\bar{q}}^{AB}. \tag{1}$$

The real part of a cross section determines the orientation transfer in a collision of particles, the relaxation, and the formation of higher polarization moments (alignment, hyperfine polarization) (Dmitriev et al., 1994). The imaginary part of a cross section determines the shifts of a magnetic resonance frequency in the system of both Zeeman and hyperfine levels of atoms. Consequently, knowledge of a spin-exchange complex cross section allows one to completely describe the processes of spin exchange collisions. The complex cross section of the spin exchange can be conventionally represented in terms of the scattering matrix:

$$q^{AB} = \frac{\pi}{k_{AB}^2} \sum_{l=0}^{\infty} (2l+1) \cdot \left[1 - T_0^{AB}(l) \cdot T_1^{AB}(l)^*\right]. \tag{2}$$

Here, $k_{AB}^2 = \mu_{AB} \cdot v_{AB} / \hbar$ is the wave vector, μ_{AB} is the reduced mass of colliding particles A and B, v_{AB} is the mean relative velocity of colliding atoms, and the asterisk * denotes complex conjugation. The scattering matrix can be represented in terms of scattering phases ($\delta_S^{AB}(l)$) in a channel with total spin S as follows:

$$T_S^{AB}(l) = \exp(2i\delta_S^{AB}(l)), \tag{3}$$

where l is the orbital quantum number.

From expression (4), it follows that the real and imaginary parts of a complex cross section have the form:

$$\bar{q}^{AB} = \frac{\pi}{k_{AB}^2} \sum_{l=0}^{\infty} (2l+1) \sin^2\left[\delta_1^{AB}(l) - \delta_0^{AB}(l)\right], \tag{4}$$

$$\bar{\bar{q}}^{AB} = \frac{\pi}{k_{AB}^2} \sum_{l=0}^{\infty} (2l+1) \sin 2\left[\delta_1^{AB}(l) - \delta_0^{AB}(l)\right]. \tag{5}$$

Thus, to calculate the cross sections of interest to us, it is necessary to calculate the scattering phases for the singlet and triple terms.

According to formulas (4) and (5), the real and imaginary parts of a complex cross section of the spin exchange (1) can be expressed in terms of scattering phases for the singlet and triplet terms of the alkali molecule. The scattering phases were determined in the Jeffrey quasi-classical approximation modified by Langer (Mott and Massey, 1965). The use of the quasi-classical approximation in calculating the scattering phases is quite justified because, in the case of alkali-metal dimers with large reduced mass μ_{AB}, the centrifugal barrier $\left(\frac{(l+1/2)^2}{2\mu_{AB}R^2}\right)$ changes slowly with increasing orbital quantum number l as compared to the kinetic energy. As a result, it is necessary to take into account the contributions of a large number of partial waves to cross sections (4) and (5).

Spin-Exchange Frequency Shifts for a Mixture of Alkali Atoms

Spin-exchange shifts of the magnetic-resonance frequency in a mixture of alkali atoms were considered in (Okunevich, 1995). Evolution of the density matrix of atoms A can be described by the following equation:

$$\frac{d}{dt}\rho^{(A)} = \left[\left(\frac{\partial}{\partial t}\right)_{AB} + \left(\frac{\partial}{\partial t}\right)_{AA} + \left(\frac{\partial}{\partial t}\right)_{AC} + \left(\frac{\partial}{\partial t}\right)_{W} + \left(\frac{\partial}{\partial t}\right)_{H}\right]\rho^{(A)}. \quad (6)$$

Here, the first term describes a change in the density matrix due to collisions of atoms A and B, the second term describes a change in the density matrix due to collisions of atoms A and A, the third term describes a change in the density matrix because of collisions of atoms A and buffer-gas atoms C, the fourth term corresponds to the influence of diffusion to the absorption-chamber walls, and the last term corresponds to the influence of interaction between atoms and dc magnetic field. If, for example, experiments are carried out in absorption chambers with an anti-relaxation coating without buffer gas (Aleksandrov, 2000), the third and fourth terms can be neglected.

The equation describing the evolution of the density matrix as a result of the spin exchange (6) should necessarily take into account the contribution from the depolarization of alkali atoms on the wall of the absorption chamber and in the volume upon collisions with buffer gas atoms, when the rate of the spin exchange is comparable with the relaxation rates caused by the diffusion (γ_D) and collisions with buffer gas atoms (γ_{AC}) (Okunevich, 1995). Actually, the diffusion rate is determined as:

$$\gamma_D = \frac{D}{\lambda_D^2}, \quad (7)$$

where D is the diffusion coefficient of atoms in the buffer gas (for example, molecular nitrogen), λ_D is the diffusion length determined from linear dimensions of the absorption chamber (the chamber used in (Dmitriev and Dovator, 2007) was 4 cm in diameter and 6 cm in length, which yields $\lambda_D =$

1.71 cm^{-2} at a buffer gas pressure of 100 Torr). At the same time, the presence of the buffer gas in the absorption chamber leads also to the destruction of the polarization of an alkali atom upon collisions. However, as follows from (Wagshul and Chupp, 1994), the joint influence of the diffusion and depolarization upon collisions with the buffer gas leads to depolarization rates much lower than the spin exchange rates $\overline{\gamma}_{AA}$ and $\overline{\gamma}_{AB}$.

In accordance with (Okunevich, 1995), consideration of only spin exchange collisions results in the following expressions for the magnetic-resonance frequency shifts of two hyperfine states of an alkali atom:

$$\Delta\omega(\pm) = \overset{(1)}{\delta}\omega(\pm) + \overset{(2)}{\delta}\omega \qquad (8)$$

Here, the first term is caused by the occurrence of an additive to the transverse orientation component of atom A in collisions with longitudinally oriented atoms B, while the second term is due to the transfer of transverse orientation from one sublevel F of atom A to another upon collisions. In correspondence with (Okunevich, 1995), the shifts in (8) can be written as:

$$\overset{(1)}{\delta}\omega(+) = -\frac{P_z(B)}{2(2I_A+1)} \cdot \left[\overline{\overline{\gamma}}_{AB} - \overline{\overline{\gamma}}_{AA} B_- \left(\frac{2I_A-1}{2I_A+1} \right)^{1/2} \right], \qquad (9)$$

$$\overset{(1)}{\delta}\omega(-) = -\frac{P_z(B)}{2(2I_A+1)} \cdot \left[\overline{\overline{\gamma}}_{AB} + \overline{\overline{\gamma}}_{AA} B_+ \left(\frac{2I_A+3}{2I_A+1} \right)^{1/2} \right], \qquad (10)$$

$$\overset{(2)}{\delta}\omega = -\frac{C}{\overline{\overline{\omega}}_0} \left\{ \left(2\overline{\gamma}_{AA} + 3\overline{\gamma}_{AB} \right)^2 - \left[\overline{\overline{\gamma}}_{AA} P_z(B) \right]^2 \right\}. \qquad (11)$$

Here, $\delta^{(1)}\omega(+)$ is the magnetic-resonance frequencyshift for the hyperfine state $F = S + I$ (S is the electronspin, which is 1/2 for alkali atoms, and I is the alkaliatomnuclear spin); $\delta^{(1)}\omega(-)$ is the magnetic-resonancefrequency shift for the hyperfine state $F = S - I$; $P_z(B)$ is the polarization of particle B; I_A is the nuclear spin of particle A; and $\overline{\overline{\gamma}}_{AB}$ and $\overline{\overline{\gamma}}_{AA}$ are the imaginary parts of the complex spin-exchange rate $\overline{\overline{\gamma}}$, which can be expressed in terms of the complex spin exchange cross section; and $\overline{\omega_0} = H_0|g_s|\mu_B/\hbar$ is the electron-precession frequency (H_0 is the dc magneticfield, g_s is the electron g factor, and μ_B is the Bohr magneton). The imaginary parts of the complex spin exchange rate can be presented as $\overline{\overline{\gamma}}_{AB} = \langle v_{AB} \rangle N_B \overline{\overline{q}}_{AB}$, where N_B is the concentration of particles B, $<v_{AB}> = (8k_BT/\pi\mu_{AB})$ is the average relative thermal velocity of the colliding particles (k_B is the Boltzmann constant, T is temperature, and μ_{AB} is the reduced mass), and is the imaginary part of the spin-exchange cross section of the colliding particles. Subscripts AA and AB correspond to collisions between identical and different alkali atoms, respectively in correspondence with (Okunevich, 1995), parameters C and $B_\pm$ have the form:

$$B_+ = \frac{2I_A + 2}{6}\left(\frac{2I_A + 3}{2I_A + 1}\right)^{1/2}, \tag{12}$$

$$B_- = \frac{2I_A}{6}\left(\frac{2I_A - 1}{2I_A + 1}\right)^{1/2}, \tag{13}$$

$$C = \frac{2I_A(2I_A + 2)(2I_A + 3)(2I_A - 1)}{288(2I_A + 1)^4}, \tag{14}$$

Table 1 lists the coefficients included in the formulas (9), (10) and (11) for various values of the nuclear spin of alkali atoms I_A.

Table 1. Values of coefficients and parameters included in expressions (9), (10) and (11), depending on I_A

	$I_A = 7/2$	$I_A = 5/2$	$I_A = 3/2$
B_+	$\dfrac{3}{2}\sqrt{\dfrac{5}{4}}$	$\dfrac{7}{6}\sqrt{\dfrac{4}{3}}$	$\dfrac{5}{6}\sqrt{\dfrac{3}{2}}$
B_-	$\dfrac{7}{6}\sqrt{\dfrac{3}{4}}$	$\dfrac{5}{6}\sqrt{\dfrac{2}{3}}$	$\dfrac{1}{2}\sqrt{\dfrac{1}{2}}$
$\left(\dfrac{2I_A - 1}{2I_A + 1}\right)^{1/2}$	$\sqrt{\dfrac{3}{4}}$	$\sqrt{\dfrac{2}{3}}$	$\sqrt{\dfrac{1}{2}}$
$\left(\dfrac{2I_A + 3}{2I_A + 1}\right)^{1/2}$	$\sqrt{\dfrac{5}{4}}$	$\sqrt{\dfrac{4}{3}}$	$\sqrt{\dfrac{3}{2}}$
C	0,0032	0,0030	0,0012

Cs-K System

In order to calculate frequency shifts of magnetic resonance lines one should know the complex spin-exchange cross sections (first of all, their imaginary parts) for the Cs–Cs, Cs–K, and K–K pairs of colliding atoms.

Cesium atom (^{133}Cs) in the ground $6s^5p^6$ state has electron spin $S = 1/2$ and nuclear spin $I = 7/2$. Thus, a cesium atom has two hyperfine levels in the ground state: $F = 3$ and $F = 4$. The complex spin exchange cross sections for these atoms were calculated in (Dmitriev et al., 2015) based on the interaction potentials presented in (Xie at al., 2009; Amiot and Dulieu, 2002).

The complex spin exchange cross sections for cesium and potassium atoms were calculated in (Kartoshkin, 2012). The calculations were carried out with the interaction potentials obtained in (Ferber et al., 2008; Ferber et al., 2009). There are two potassium isotopes in nature: ^{39}K and ^{41}K. Atoms of the ^{39}K isotope in the ground $4s^3p^6$ state with nuclear spin $I = 3/2$ and two hyperfine states $F = 2$ and $F = 1$ and atoms ^{133}Cs in the ground $6s^5p^6$ state with electron spin $S = 1/2$ and nuclear spin $I = 7/2$ were used in the Cs–K tandem magnetometer.

The complex spin exchange cross sections for the ^{39}K–^{39}K atomic pair were calculated in (Kartoshkin, 2011, 881-884) based on the interaction potentials presented in (Amiot, 1991; Amiot et al., 1995; Ahmed et al., 2005).

The frequency shifts were calculated based on expressions (9) and (10) taking into account the previously obtained temperature dependences of the imaginary parts of the complex spin-exchange cross sections (Dmitriev et al., 2015; Kartoshkin, 2012; Kartoshkin, 2011, 881-884). The values of the coefficients entering (9) and (10) for Cs or K atom used as atom A are listed in Table 1.

Since expressions (9), (10) include the temperature dependences of the imaginary parts of the complex spin exchange rate, it is necessary to know the concentration of alkaline atoms in the absorption chamber to calculate the values of shifts. The tables from (Nesmeyanov, 1963, 367) were used to calculate the concentration values. Since the case of a mixture of alkaline atoms is considered in this paper, it is necessary to use Raoul's law for the pressure of saturated vapor over the melt of a mixture of metals to transition from the temperature of the absorption chamber to the concentration of alkaline atoms in it.

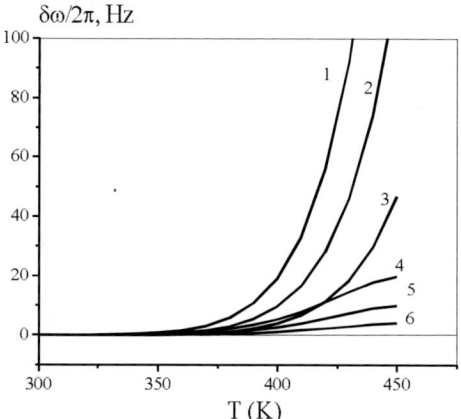

Figure 1. Temperature dependences of the magnetic resonance frequency shift of the ^{39}K atoms at the optical pumping of Cs atoms (based on (Kartoshkin, 2020, 1355-1358) data).

1 - $\delta^{(1)}\omega(+)/2\pi$, $P_z(B) = 100\%$,
2 - $\delta^{(1)}\omega(+)/2\pi$, $P_z(B) = 50\%$,
3 - $\delta^{(1)}\omega(+)/2\pi$, $P_z(B) = 20\%$,
4 - $\delta^{(1)}\omega(-)/2\pi$, $P_z(B) = 100\%$,
5 - $\delta^{(1)}\omega(-)/2\pi$, $P_z(B) = 50\%$,
6 - $\delta^{(1)}\omega(-)/2\pi$, $P_z(B) = 20\%$.

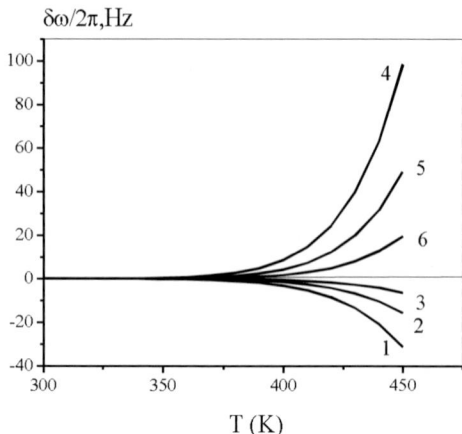

Figure 2. Temperature dependences of the magnetic resonance frequency shift of the ^{133}Cs atoms at the optical pumping of ^{39}K atoms (based on (Kartoshkin, 2020, 1355-1358) data).

1 - $\delta^{(1)}\omega(+)/2\pi$, $P_z(B) = 100\%$,
2 - $\delta^{(1)}\omega(+)/2\pi$, $P_z(B) = 50\%$,
3 - $\delta^{(1)}\omega(+)/2\pi$, $P_z(B) = 20\%$,
4 - $\delta^{(1)}\omega(-)/2\pi$, $P_z(B) = 100\%$,
5 - $\delta^{(1)}\omega(-)/2\pi$, $P_z(B) = 50\%$,
6 - $\delta^{(1)}\omega(-)/2\pi$, $P_z(B) = 20\%$.

The calculation results are shown in Figures 1 and 2. As one can see from Figure 1, both frequency shifts ($\delta^{(1)}\omega(+)$ and $\delta^{(1)}\omega(-)$) increase with an increase in the temperature in the absorption chamber (and, accordingly, the alkali atomic concentration). Figure 1 shows the temperature dependences of the magnetic resonance frequency shifts of K atoms at the optical orientation of Cs (here it is atom B) atoms.

This behavior of the frequency shift is due to the competition between the contribution to the shift value (formula (10)) of terms due to collisions between identical atoms - AA and different - AB, which in turn depends on the concentrations of the corresponding atoms, and on the temperature dependences of the imaginary parts of the complex cross sections spin exchange. In addition, the shift value in accordance with formula (10) is determined by the coefficients presented in Table 1.

Figure 2 shows the temperature dependences of the magnetic resonance frequency shifts of Cs atoms at the optical orientation of K atoms. As follows from Figure 2, at the optical orientation of K atoms, shifts ($\delta^{(1)}\omega(+)$ and $\delta^{(1)}\omega(-)$) have different signs. Note that shift $\delta^{(1)}\omega(+)$ induced by collisions between Cs and optical pumped K atoms is negative in the entire temperature range under study, whereas other shift ($\delta^{(1)}\omega(-)$) is positive in the entire temperature range under study.

The magnetic-resonance frequency shift $\delta^{(2)}\omega$ corresponding to the second term in (8) is much smaller than shifts $\delta^{(1)}\omega(\pm)$. Indeed, in a magnetic field of about 10^{-4} T, the value of $\overline{\omega_0} = H_0 |g_s| \mu_B / \hbar$ is ~10^5 Hz and the value in the braces in expression (11) is ~10^6 s^{-1} at the atomic concentration of about 10^{12} cm^{-3}. Taking into account the C value from Table 1, we find that $\delta^{(2)}\omega/2\pi$ is about 10^{-4} Hz (i.e., much smaller than the $\delta^{(1)}\omega(\pm)$ value), and the shifts of the second type can be neglected. Therefore, in the future, this shift will not be considered and taken into account for other systems under study.

Figures 1 and 2 show the calculation results for three polarizations of atoms B ($P_z(B) = 100$, 50 and 20%).

K-Rb System

In the collision of the ^{39}K and Rb alkali atoms in the ground state with an electron spin $S = 1/2$ a molecule of KRb is formed. This molecule is described by two potentials corresponding to the total spins of the system $S= 0$ (singlet term) and $S = 1$ (triplet term). In (Amiot, 1991; Amiot et al., 1995; Ahmed et al., 2005), interaction potentials describing the ^{39}K$_2$ dimers were presented. In (Jenc and Brandt, 1989; Jenc, 1983) the potentials for the ^{85}Rb$_2$ were presented, and in (Pashov, 2007) the potentials were received for the ^{39}K^{85}Rb dimer. Based on the presented interaction potentials, in (Kartoshkin, 2011, 665-671; Kartoshkin, 2015; Kartoshkin, 2011, 881-884), the real and imaginary parts of the complex spin exchange cross sections were calculated for the systems under study. The calculations were carried out in accordance with expressions (9) and (10).

As shown above the magnitude of the shift determined by relation (11) is significantly less than the magnitudes of the shifts (9) and (10). In the future, we will only consider these shifts. Thus, in order to calculate the shifts of the magnetic resonance frequency that are of interest to us due to the spin exchange, in accordance with (9) and (10), we need to know the values of the complex cross sections of the spin exchange (their imaginary parts) for the following pairs of atoms: Rb-Rb, Rb-K and K-K. To calculate the frequency shifts, it is necessary to know the concentration of alkali atoms in a mixture at different temperatures.

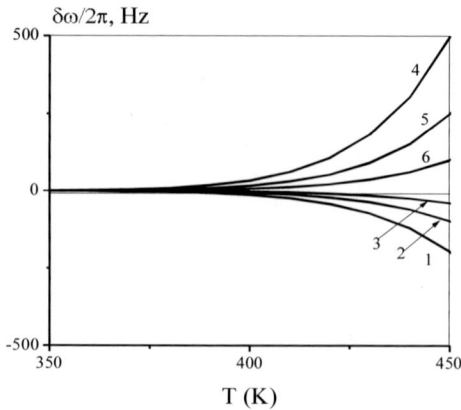

Figure 3. Temperature dependences of the magnetic resonance frequency shift of the ^{85}Rb atoms at the optical pumping of ^{39}K atoms (based on (Kartoshkin, 2020, 012146) data).

1 - $\delta^{(1)}\omega(+)/2\pi$, $P_z(B) = 100\%$,
2 - $\delta^{(1)}\omega(+)/2\pi$, $P_z(B) = 50\%$,
3 - $\delta^{(1)}\omega(+)/2\pi$, $P_z(B) = 20\%$,

4 - $\delta^{(1)}\omega(-)/2\pi$, $P_z(B) = 100\%$,
5 - $\delta^{(1)}\omega(-)/2\pi$, $P_z(B) = 50\%$,
6 - $\delta^{(1)}\omega(-)/2\pi$, $P_z(B) = 20\%$.

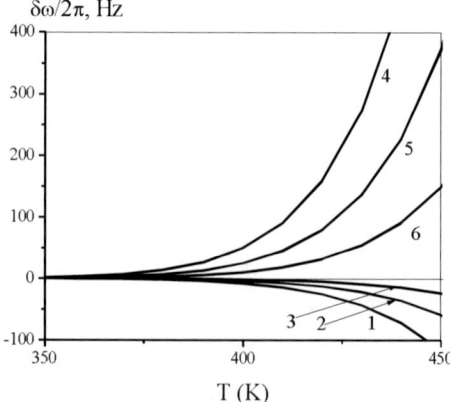

Figure 4. Temperature dependences of the magnetic resonance frequency shift of the ^{87}Rb atoms at the optical pumping of ^{39}K atoms (based on (Kartoshkin, 2020, 012146) data).

1 - $\delta^{(1)}\omega(+)/2\pi$, $P_z(B) = 100\%$,
2 - $\delta^{(1)}\omega(+)/2\pi$, $P_z(B) = 50\%$,
3 - $\delta^{(1)}\omega(+)/2\pi$, $P_z(B) = 20\%$,

4 - $\delta^{(1)}\omega(-)/2\pi$, $P_z(B) = 100\%$,
5 - $\delta^{(1)}\omega(-)/2\pi$, $P_z(B) = 50\%$,
6 - $\delta^{(1)}\omega(-)/2\pi$, $P_z(B) = 20\%$.

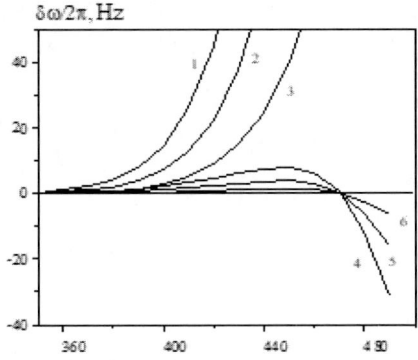

Figure 5. Temperature dependences of the magnetic resonance frequency shift of the ^{39}K atoms at the optical pumping of Rb atoms.

1 - $\delta^{(1)}\omega(+)/2\pi$, $P_z(B) = 100\%$,
2 - $\delta^{(1)}\omega(+)/2\pi$, $P_z(B) = 50\%$,
3 - $\delta^{(1)}\omega(+)/2\pi$, $P_z(B) = 20\%$,
4 - $\delta^{(1)}\omega(-)/2\pi$, $P_z(B) = 100\%$,
5 - $\delta^{(1)}\omega(-)/2\pi$, $P_z(B) = 50\%$,
6 - $\delta^{(1)}\omega(-)/2\pi$, $P_z(B) = 20\%$.

Figures 3-5 present the results of calculating the magnetic resonance frequency shifts of the ^{39}K, ^{85}Rb, and ^{87}Rb atoms in case of the optical pumping of K or Rb atoms.

Figures 3 and 4 show the frequency shifts of the magnetic resonance of ^{85}Rb and ^{87}Rb at optical pumping of ^{39}K atoms. Magnetic resonance shifts ($\delta^{(1)}\omega(+)$ and $\delta^{(1)}\omega(-)$) have different signs for the two hyperfine states. For the hyperfine state **F = I+S**, the frequency shift is negative, while for the hyperfine state **F = I-S** it is positive.

Figure 5 shows the temperature dependence of the magnetic resonance shift of potassium atoms upon optical pumping of rubidium atoms. As follows from Figure 5, the shift value ($\delta^{(1)}\omega(+)$) increases with increasing temperature, being all the time in the region of positive values. At the same time, the value of the shift ($\delta^{(1)}\omega(-)$) with increasing temperature first increases to a temperature of the order of 450 K being in the region of positive values of the shift, and then begins to decrease in value, passing in the region of T=470 K through zero and further increases in value in the region of negative values. This behavior of the frequency shift is due to the competition between the contribution to the shift value (formula (10)) of terms due to collisions between identical atoms - AA and different - AB, which in turn depend on the concentrations of the corresponding atoms, and on the temperature dependences of the imaginary parts of the complex cross sections spin exchange. In addition, the shift value in accordance with formulas (9) and (10) is determined by the coefficients presented in Table 1.

Na-K System

To calculate the magnetic resonance frequency shift in accordance with formulas (9) and (10), it is necessary to determine the spin-exchange rate constants, which depend on the velocities of colliding particles, the corresponding cross sections, and the concentrations of colliding atoms ($\gamma_{AB} = <v_{AB}>N_B q_{AB}$). The concentration of alkali atoms was determined in accordance with the data reported in (Nesmeyanov, 1963, 367). Since the mixture of alkali atoms is used in the experiment, it is necessary to use the Raul law for the saturated vapor pressure above the melt of the mixture of metals to pass from the temperature in the absorption chamber to the concentration of alkali atoms in it. In the calculations, it was assumed that the potassium and sodium alkali metals are present in the absorption chamber in equal weight parts.

Further, the temperature dependences of shifts $\delta^{(1)}\omega(+)$ and $\delta^{(1)}\omega(-)$ were calculated in accordance with expressions (9) and (10). Figure 6 shows thetemperature dependences of shifts for different degrees of polarization of K atoms (quantity P_z), calculated using the interaction potentials from (Ivanov and Sovkov, 2003; Zemke and Stwalley, 1994) for dimer Na_2 and from (Gerdes, 2008) for dimer NaK. To calculate the complex spin exchange cross sections at the collision of two K atoms, we used the interaction potentials of the singlet and triplet terms presented in (Amiot, 1991; Amiot et al.; 1995, Ahmed et al., 2005).

The complex spin exchange cross sections for collision of two Na atoms were calculated in (Kartoshkin, 2014) using the interaction potentials from (Ivanov and Sovkov, 2003; Zemke and Stwalley, 1994) and for the Na–K system in (Kartoshkin, 2010) with account for the interaction potentials from (Gerdes, 2008). The complex spin exchange cross sections for collision of two K atoms were calculated in (Kartoshkin, 2011, 881-884) using the interaction potentials from (Amiot, 1991; Amiot et al., 1995; Ahmed et al., 2005).

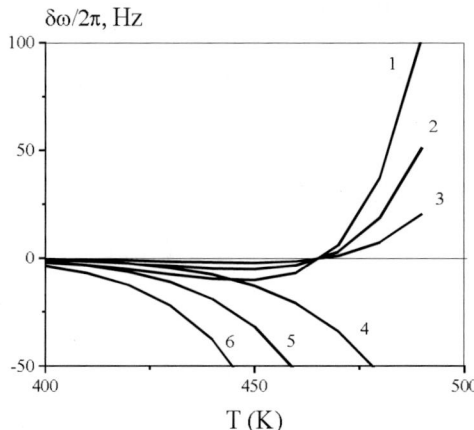

Figure 6. Temperature dependences (Kartoshkin, 2021, 1221-1227) of the magnetic resonance frequency shifts for sodium atoms in the ground state due to the spin-exchange in the Na–K mixture with potentials from (Ivanov and Sovkov, 2003; Zemke and Stwalley, 1994) for the Na–Na pair and from (Gerdes, 2008) for the Na–K pair:

1 - $\delta^{(1)}\omega(-)/2\pi$, $P_z(B) = 100\%$,
2 - $\delta^{(1)}\omega(-)/2\pi$, $P_z(B) = 50\%$,
3 - $\delta^{(1)}\omega(-)/2\pi$, $P_z(B) = 20\%$,
4 - $\delta^{(1)}\omega(+)/2\pi$, $P_z(B) = 20\%$,
5 - $\delta^{(1)}\omega(+)/2\pi$, $P_z(B) = 50\%$,
6 - $\delta^{(1)}\omega(-)/2\pi$, $P_z(B) = 100\%$.

As a rule, the polarization of atoms in experiments on the optical orientation of alkali atoms reaches several tens of percent. The curves shown in Figure 6 are plotted for P_z=100, 50, and 20%. It can be seen from the figure that, in the temperature interval under investigation, shifts $\delta^{(1)}\omega(-)$ increase in magnitude towards positive values, while shifts $\delta^{(1)}\omega(+)$ increase in magnitude and remain negative in the entire temperature range. If, however, we plot the curves shown in Figure 6 on a smaller scale, the following behavior of shifts $\delta^{(1)}\omega(-)$ and $\delta^{(1)}\omega(+)$ near $T = 450$ K can be observed. The value of shifts $\delta^{(1)}\omega(+)$ is negative in the entire temperature range, while $\delta^{(1)}\omega(-)$ increases in absolute value, remaining negative down to temperature near 430 K. Then the shift begins decreasing in the absolute value, but remains negative to the temperature range of 440–460 K. In the temperature interval of 460–470 K, the frequency shift passes through zero and begins increasing in magnitude, remaining positive.

Figure 6 shows the dependences of the frequency shifts under investigation on a smaller scale, which permits the observation of the detected feature. Figure 7 shows analogous dependences for the case when potentials from (Zemke and Stwalley, 1994) are used in calculations for dimer Na$_2$. In this case, the transition through zero occurs in the temperature interval of 440–450 K; i.e., the point of polarity reversal of frequency shift $\delta^{(1)}\omega(-)$ is displaced towards lower temperatures. Figure 7 shows the temperature dependences of the magnetic resonance frequency shift for sodium atoms in the ground state caused by the spin exchange in the Na–K mixture in the temperature interval of 350–450 K on an enlarged scale (Kartoshkin, 2021, 1221-1227).

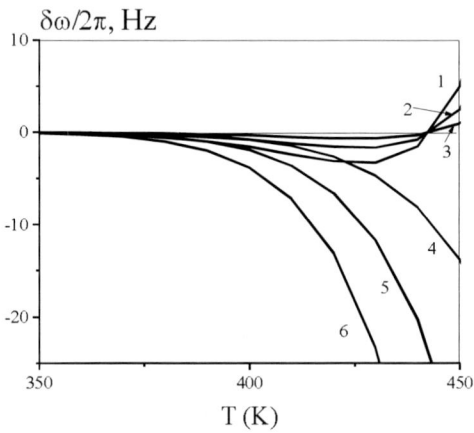

Figure 7. Temperature dependences (Kartoshkin, 2021, 1221-1227) of the magnetic resonance frequency shifts for sodium atoms in the ground state caused by the spin exchange in the Na–K mixture for potentials from (Zemke and Stwalley, 1994) and for the Na–Na pair and with potentials from (Gerdes, 2008) and for the Na–K pair:

1 - $\delta^{(1)}\omega(-)/2\pi$, $P_z(B) = 100\%$,
2 - $\delta^{(1)}\omega(-)/2\pi$, $P_z(B) = 50\%$,
3 - $\delta^{(1)}\omega(-)/2\pi$, $P_z(B) = 20\%$,
4 - $\delta^{(1)}\omega(+)/2\pi$, $P_z(B) = 20\%$,
5 - $\delta^{(1)}\omega(+)/2\pi$, $P_z(B) = 50\%$,
6 - $\delta^{(1)}\omega(-)/2\pi$, $P_z(B) = 100\%$.

The following remarks can be made in conclusion. In the calculation of the magnetic resonance frequency shifts for other pairs of alkali atoms (e.g., K–Cs (Kartoshkin, 2020,1355-1358), no sign reversal in the temperature dependence of the frequency shifts were observed for quantity $\delta^{(1)}\omega(+)$ as well as $\delta^{(1)}\omega(-)$ in the entire temperature range. However, vanishing of the magnetic resonance frequency shift and passage through zero of cesium atoms in the Cs–Rb mixture has been observed earlier in (Okunevich, 1995). The presence of a buffer gas was taken into account. As a result, vanishing of shift

$\delta^{(1)}\omega(-)$ was observed upon a change in the buffer gas pressure in the absorption chamber, and for $\delta^{(1)}\omega(+)$, it was observed upon a change in the absorption chamber temperature in the presence of a buffer gas.

The sign reversal in magnetic resonance frequency shift $\delta^{(1)}\omega(-)$ observed in this study is due to the competition between two processes. It follows from formula (10) that the contribution to the frequency shift comes from two spin-exchange processes competing with each other: (i) the collision of sodium atoms with potassium atoms (the contribution from this process is determined by the first term in the brackets in expression (10)) and (ii) collisions between sodium atoms (second term in the brackets in expression (10)). In turn, these terms are determined by quantities $\bar{\bar{\gamma}}_{AB}$ and $\bar{\bar{\gamma}}_{AA}$ (where $\bar{\bar{\gamma}}_{AB} = \langle v_{AB} \rangle N_B \bar{\bar{q}}_{AB}$ and $\bar{\bar{\gamma}}_{AA} = \langle v_{AA} \rangle N_A \bar{\bar{q}}_{AA}$). Here, subscript AA corresponds to collisions between sodium atoms and subscript AB, to collisions between sodium and potassium atoms. As follows from the data represented in (Kartoshkin, 2021), the imaginary part of the spin exchange cross section is negative in the entire range of temperatures considered here for collisions between sodium atoms, while the imaginary part of the spin exchange cross section for collisions between sodium and potassium atoms is positive. With increasing temperature in the absorption chamber, the concentrations of sodium and potassium atoms increase, and the spin-exchange cross sections change. Therefore, magnetic resonance frequency shift $\delta^{(1)}\omega(-)$ in the range of lower temperatures is determined by the second term in formula (10) (depending on the spin exchange between sodium atoms). This term is negative. With increasing temperature, the role of the first term in equation (10), which is positive, gradually increases (it is determined by the spin exchange between sodium and potassium atoms). It is the competition between these two terms (with the first term increasing faster) that determines the sign reversal (from negative to positive) in the temperature dependence, and the temperature dependence of the magnetic resonance frequency shift $\delta^{(1)}\omega(-)$ passes through zero.

At the same time, the quantity in the brackets in formula (9) is positive in the entire temperature range under investigation because, according to (Kartoshkin, 2021, 1221-1227), the imaginary parts of complex cross sections (1) for Na–Na and Na–K pairs have different signs (negative for two sodium atoms and positive for sodium and potassium atoms), with the contribution from the spin exchange between two sodium atoms appearing with a minus sign. Therefore, the square bracket in expression (9) has a positive sign, and, taking into account the minus sign of this bracket in expression (9), the magnetic resonance frequency shift $\delta^{(1)}\omega(+)$ has a negative sign and increases in magnitude with temperature.

It should be noted that the development of all quantum electronics devices operating on the principle of optical orientation of atoms (i.e., quantum magnetometers, frequency standards, quantum gyroscopes, and quantum magnetic encephalographs based on quantum magnetometers) encounters a problem associated with the magnetic resonance frequency shift. This is due to the fact, that the magnetic resonance frequency is measured in these devices. The true value of the quantity being measured can be distorted by the frequency shift caused by various physical processes (spin exchange, the effect of pump radiation, collisions with buffer-gas atoms, and so on). Thus, if the possible behavior is known of the magnetic resonance frequency shift (which in our case depends on the spin exchange during collisions of polarized alkali atoms), it is possible to select the conditions (as well as the appropriate mixtures of alkali atoms) for minimizing the negative effect of the frequency shift on the precision characteristics of devices.

K-Li System

Let us consider a situation when the optical orientation of K atoms is carried out in a mixture of alkaline K and Li atoms. In accordance with expressions (9) and (10), we need to know the temperature dependences of the imaginary parts of the complex spin exchange rate (and $\overline{\overline{\gamma}}_{AB} = \langle v_{AB} \rangle N_B \overline{\overline{q}}_{AB}$ and $\overline{\overline{\gamma}}_{AA} = \langle v_{AA} \rangle N_A \overline{\overline{q}}_{AA}$) in the case when atom A is a potassium atom and atom B is a litium atom. In (Kartoshkin, 2011, 881-884; Kartoshkin, 2019; Kartoshkin, 2021, 641-644), the temperature dependences of the real and imaginary parts of the complex cross section of the spin exchange for K-K, K-Li and Li-Li pairs were calculated. As follows from (Kartoshkin, 2011, 881-884; Kartoshkin, 2021, 641-644)., the imaginary parts of the spin exchange cross sections for pairs of identical K-K and Li-Li atoms are positive in magnitude and decrease with increasing temperature. At the same time, the imaginary part of the spin exchange cross section for the K-Li pair increases in magnitude and remains positive throughout the temperature range (Kartoshkin, 2019).

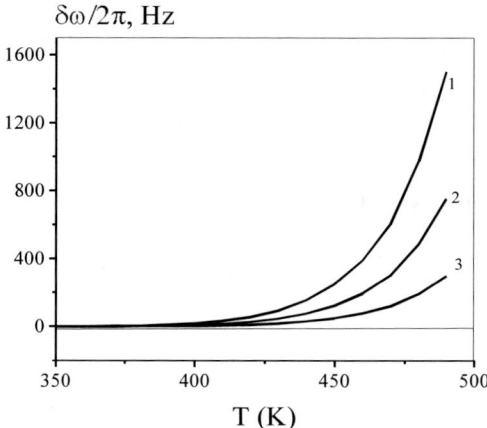

Figure 8. Temperature dependences of the magnetic resonance frequency shift of the ^{7}Li atoms at the optical pumping of ^{39}K atoms in the Li–K mixture (Kartoshkin, 2022).

1 – $\delta^{(1)}\omega$ (+)/2π и $\delta^{(1)}\omega$ (-)/2π, P = 100%,
2 – $\delta^{(1)}\omega$ (+)/2π и $\delta^{(1)}\omega$ (-)/2π, P = 50%,
3 – $\delta^{(1)}\omega$ (+)/2π и $\delta^{(1)}\omega$ (-)/2π, P = 20%.

Figures 8 present the results of calculating the magnetic resonance frequency shifts of the ^{7}Li atoms in case of the optical pumping of ^{39}K atoms.

The calculation was carried out in accordance with formulas (9)-(10) using data on imaginary parts of complex spin exchange cross sections presented in (Kartoshkin, 2011, 881-884, Kartoshkin,2019; Kartoshkin, 2021, 641-644). Since expressions (9), (10) include the temperature dependences of the imaginary parts of the complex spin exchange rate, it is necessary to know the concentration of alkaline atoms in the absorption chamber to calculate the values of shifts. The tables from (Nesmeyanov, 1963, 367) were used to calculate the concentration values. Since the case of a mixture of alkaline atoms is considered in this paper, it is necessary to use Raoul's law for the pressure of saturated vapor over the melt of a mixture of metals to transition from the temperature of the absorption chamber to the concentration of alkaline atoms in it.

As can be seen from Figure 8, the shifts of $\delta^{(1)}\omega(+)$ and $\delta^{(1)}\omega(-)$ Li atoms coincide in magnitude and grow throughout the range of temperatures studied. The coincidence of the shift values is due to the fact that the imaginary parts of the complex spin exchange rate ($\bar{\bar{\gamma}}_{AB} = \langle v_{AB} \rangle N_B \bar{\bar{q}}_{AB}$ and $\bar{\bar{\gamma}}_{AA} = \langle v_{AA} \rangle N_A \bar{\bar{q}}_{AA}$))

included in expressions (9) and (10) depend on the concentration of alkaline atoms, the corresponding cross section and the relative velocity of colliding atoms. Moreover, if the collision velocities and cross-sections are close in magnitude, then the concentrations of Li and K atoms differ by several orders of magnitude, and the concentration of K atoms is higher. In addition, the shift value in accordance with formulas (9) and (10) is determined by the coefficients presented in Table 1. Thus, in expressions (9) and (10), the first terms in parentheses prevail, which depend on the concentration of K atoms and determine both the magnitude and the sign of the magnetic resonance frequency shift.

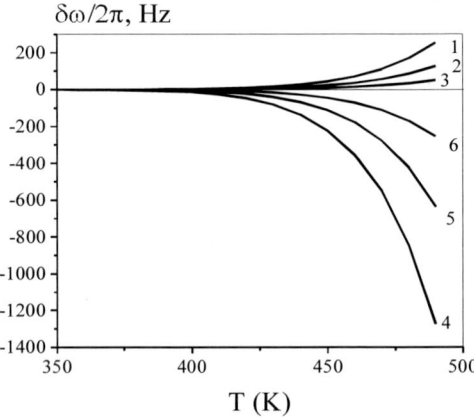

Figure 9. Temperature dependences of the magnetic resonance frequency shift of the ^{39}K atoms at the optical pumping of ^{7}Li atoms in the Li–K mixture (Kartoshkin, 2022).

1 – $\delta^{(1)}\omega$ (+)/2π, P_z = 100%,
2 – $\delta^{(1)}\omega$ (+)/2π, P_z = 50%,
3 – $\delta^{(1)}\omega$ (+)/2π, P_z = 20%,

4 – $\delta^{(1)}\omega$ (−)/2π, P_z = 100%,
5 – $\delta^{(1)}\omega$ (−)/2π, P_z = 50%,
6 – $\delta^{(1)}\omega$ (−)/2π, P_z = 20%.

With the optical orientation of ^{7}Li atoms in a K-Li mixture, temperature dependences of the shifts in the magnetic resonance frequency of potassium atoms for two hyperfine states **F = 2** and **F = 1** were calculated. Figure 9 shows the temperature dependences of the magnetic resonance frequency shift for the hyperfine states **F = 1** and **F = 2** of the ^{39}K atom calculated on the basis of ratios (9) and (10) and data on imaginary parts of the spin exchange cross sections for the K-K and K-Li pairs shown in (Kartoshkin, 2011, 881-884; Kartoshkin, 2019).

As follows from Figure 9, the temperature dependences of the frequency shift for the K atom in the ground state differ significantly from the dependences presented in Figure 8 for the lithium atom. This is due to the fact, that in the case of the optical orientation of the Li atoms and the observation of the shift of the K atoms, the second term in expression (10) plays an essential role, since it depends on the concentration of K atoms in the absorption chamber. As noted earlier, the concentration of K atoms in the absorption chamber is much higher than the concentration of Li atoms at the same temperatures. The difference between the dependencies $\delta^{(1)}\omega(+)$ and $\delta^{(1)}\omega(-)$ is due to the difference in the values of the coefficients B_- and B_+ (see Table 1), the explicit form of which is represented by expressions (12) and (13) and signs between the first and second terms in expression (10).

As already noted in the Introduction, magnetic resonance frequency shifts play an essential role in devices whose operating principle is based on the optical orientation of alkali metal atoms. In this case, the behavior of frequency shifts depends on the complex cross-section of the spin exchange, which is determined by the interaction potentials of atoms used in the absorption chambers as working media. The interaction potentials of alkaline atoms determine both the magnitude of the imaginary part of the complex cross section of the spin exchange and its sign. In addition, the concentrations of atomic particles in the working chamber make a significant contribution to the shifts. The influence of these factors is clearly seen in Figure 8 and Figure 9.

Cs-Rb System

Figures 10-12 show the temperature dependences of the magnetic resonance shift of cesium and rubidium atoms upon optical pumping of rubidium or cesium atoms. As follows from Figure 10 and Figure 11, the shift values $\delta^{(1)}\omega(+)$ and $\delta^{(1)}\omega(-)$ increase with increasing temperature, being all the time in the region of positive values. At the same time, the shifts $\delta^{(1)}\omega(-)$ are larger than the shifts $\delta^{(1)}\omega(+)$ in this temperature range with the same values of the degree of ^{133}Cs atoms polarization. This situation occurs for both ^{85}Rb and ^{87}Rb atoms. These behaviors of the frequency shift are due to the competition between the contribution to the shift value (formulas (9) and (10)) of terms due to collisions between identical atoms - AA and different - AB, which in turn depend on the concentrations of the corresponding atoms, and on the temperature dependences of the imaginary parts of the complex cross sections spin exchange.

Figure 12 shows the dependences of the magnetic resonance frequency shifts of cesium atoms ($\delta^{(1)}\omega(+)$ and $\delta^{(1)}\omega(-)$) on temperature for the case when the optical orientation of rubidium atoms is carried out. As follows from Figure 12, the shift values $\delta^{(1)}\omega(+)$ and $\delta^{(1)}\omega(-)$ increase with increasing temperature, being all the time in the region of positive values. At the same time, the shifts $\delta^{(1)}\omega(-)$ are larger than the shifts $\delta^{(1)}\omega(+)$ in this temperature range with the same values of the degree of Rb atoms polarization. However, in contrast to the case presented in Figures 10 and 11, the values of the shifts $\delta^{(1)}\omega(+)$ are closer in value to the shifts $\delta^{(1)}\omega(-)$ than the corresponding dependences shown in Figures 10 and 11.

As noted above, these behaviors of the frequency shift are due to the competition between the contribution to the shift value (formula (9) and (10)) of terms due to collisions between identical atoms - AA and different - AB, which in turn depend on the concentrations of the corresponding atoms, and on the temperature dependences of the imaginary parts of the complex cross sections spin exchange.

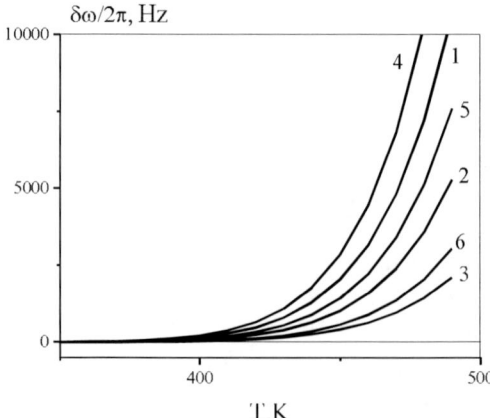

Figure 10. Temperature dependences of the magnetic resonance frequency shift of the ^{85}Rb atoms at the optical pumping of ^{133}Cs atoms in the Cs–Rb mixture.

1 – $\delta^{(1)}\omega$ (+)/2π, P_z = 100%,   4 – $\delta^{(1)}\omega$ (-)/2π, P_z = 100%,
2 – $\delta^{(1)}\omega$ (+)/2π, P_z = 50%,   5 – $\delta^{(1)}\omega$ (-)/2π, P_z = 50%,
3 – $\delta^{(1)}\omega$ (+)/2π, P_z = 20%,   6 – $\delta^{(1)}\omega$ (-)/2π, P_z = 20%.

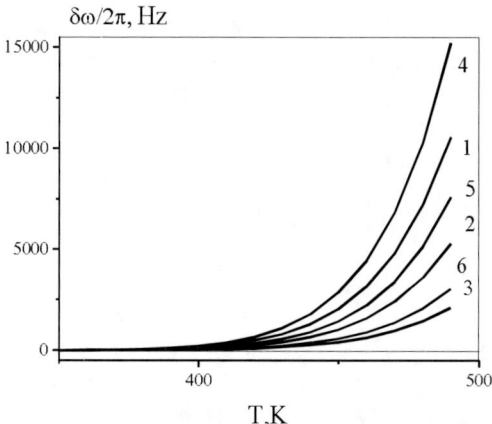

Figure 11. Temperature dependences of the magnetic resonance frequency shift of the ^{87}Rb atoms at the optical pumping of ^{133}Cs atoms in the Cs–Rb mixture.

1 – $\delta^{(1)}\omega$ (+)/2π, P_z = 100%,   4 – $\delta^{(1)}\omega$ (-)/2π, P_z = 100%,
2 – $\delta^{(1)}\omega$ (+)/2π, P_z = 50%,   5 – $\delta^{(1)}\omega$ (-)/2π, P_z = 50%,
3 – $\delta^{(1)}\omega$ (+)/2π, P_z = 20%,   6 – $\delta^{(1)}\omega$ (-)/2π, P_z = 20%.

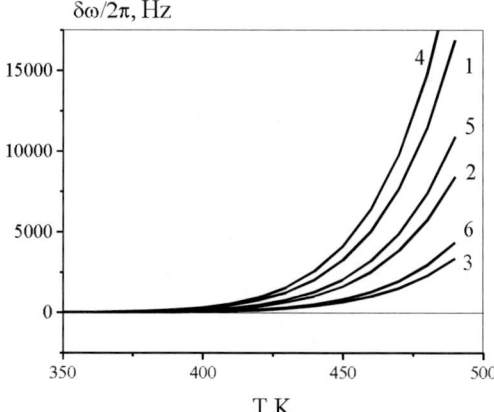

Figure 12. Temperature dependences of the magnetic resonance frequency shift of the ^{133}Cs atoms at the optical pumping of Rb atoms in the Cs–Rb mixture.

1 – $\delta^{(1)}\omega$ (+)/2π, P_z = 100%,   4 – $\delta^{(1)}\omega$ (-)/2π, P_z = 100%,
2 – $\delta^{(1)}\omega$ (+)/2π, P_z = 50%,   5 – $\delta^{(1)}\omega$ (-)/2π, P_z = 50%,
3 – $\delta^{(1)}\omega$ (+)/2π, P_z = 20%,   6 – $\delta^{(1)}\omega$ (-)/2π, P_z = 20%.

Shown in Figures 10-12, the temperature dependences of the magnetic resonance frequency shifts were obtained using data on the complex spin exchange cross sections obtained earlier in (Kartoshkin, 2016; Kartoshkin, 2015; Dmitriev et. al, 2015).

Cs-Rb System. Experimental Results

In (Dmitriev and Dovator, 2007), magnetic-resonance-frequency shifts were measured for the first time for $5s^2S_{1/2}$ rubidium atoms caused by spin-exchange collisions with optically oriented ^{133}Cs atoms. Figure 13 shows the experimental dependence of the absolute value of a doubled magnetic resonance frequency shift ($|2\delta f|$) on the temperature in an absorption chamber and the calculated dependence of the same cross sections obtained in (Kartoshkin, 2017).

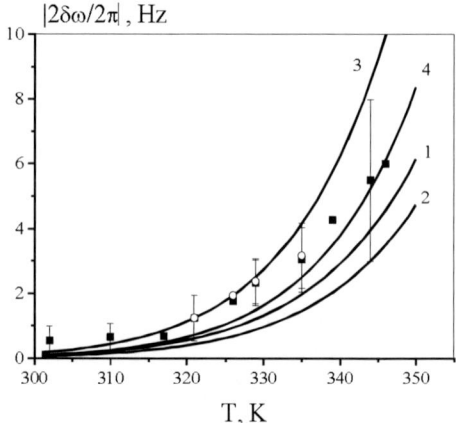

Figure 13. Temperature dependences of the absolute values of magnetic-resonance-frequency shifts for the hyperfine states of ^{87}Rb atoms. The calculated values (solid curves): (4) **F = 2**, (2) **F = 1** (Kartoshkin, 2017), (3) **F = 2**, and (1) **F = 1** (Kartoshkin, 2010, 866-869). Experimental values (symbols): (○) **F =1**, (■) **F = 2** (Dminriev and Dovator, 2007).

Taking into account, in the equations (6), the contribution associated with depolarization of alkali metal atoms on the wall of an absorption chamber and within the chamber in collisions with buffer-gas atoms is necessary when spin-exchange rate is comparable with the relaxation rate caused by diffusion γ_D and by collisions with buffer-gas atoms γ_{AC} (Okunevich, 1995).

The effects of diffusion and depolarizing collisions of Rb atoms with nitrogen molecules were considered in detail in (Kartoshkin, 2010, 866-869), where it was shown that, in calculating the magnetic resonancefrequency shifts under our conditions, it is possible to use expressions obtained without taking into account diffusion and depolarization on a buffer gas, which significantly simplifies the calculations. In experiments on optical orientation of atoms (Dmitriev and Dovator, 2007), the magnetic resonance frequency shifts of rubidium atoms were measured for two hyperfine states $\mathbf{F_A} = I_A + 1/2 = 2$ and $\mathbf{F_A = I_A - 1/2 = 1}$ (^{87}Rb isotope, $\mathbf{I_A = 3/2}$).

The absorption chamber with a mixture of equal masses of ^{87}Rb and ^{133}Cs and nitrogen, as a buffer gas, at a pressure of 100 Torr was placed in a thermostat. The concentration of alkali-metal atoms in the absorption chamber was varied by changing the thermostat temperature from 20 to 80°C. The ^{133}Cs atoms were optically oriented by resonance radiation of the D_1 cesium line. The degree of polarization of alkali metal atoms in such experiments does not exceed usually several tens of percent. Magnetic resonance was then excited at a Zeeman frequency of rubidium atoms for each of the hyperfine states, which gave rise to a change in pump-light absorption by cesium atoms due to the spin-exchange process between Cs and Rb. The change of pump-light polarization from (+) to (–) allowed a doubled magnitude of a magnetic-resonance-frequency shift to be recorded. The dependences of the absolute value of a doubled magnetic resonance-frequency shift of rubidium atoms on the temperature of an absorption chamber are shown in the figure. The concentration of alkali-metal atoms was determined using the tables presented in (Nesmeyanov, 1963, 367). Because a mixture of alkali-metal atoms was used in the experiment, it is necessary to use the Raoult law for a saturated vapor pressure above the melt of a metal mixture in order to pass from the temperature of an absorption chamber to the concentration of alkalimetal atoms in it.

To calculate the magnetic-resonance-frequency shifts defined by expressions (9) and (10), it is necessary to know the imaginary part of spin-exchange complex cross section (1). The temperature dependences of these quantities (q_{AA} and q_{AB}) for the pairs of atoms Rb–Rb and Rb–Cs were calculated in (Kartoshkin, 2015; Kartoshkin, 2016). Thus, it is possible to calculate the magnetic-resonance-frequency shifts on the basis of relationships (9) and (10) using the values of cross sections obtained in (Kartoshkin, 2015; Kartoshkin, 2016). The results of such calculation for electron polarization

$P_z(B) = 0.3$ of atom B (in our case, this is a Cs atom) are presented in Figure 13.

The calculated dependences of shift values are presented for the degree of polarization of 30%. On passage from the cross sections obtained in (Kartoshkin, 1995) to the cross sections calculated with the interaction potentials from (Kartoshkin, 2015; Kartoshkin, 2016), the difference between the shifts for the **F = 2** and **1** states decreased by a factor of more than 2. In addition, the absolute values of cross sections calculated for the **F = 2** and **1** hyperfine states became smaller than those obtained in previous calculations. One can see from the experimental data that the absolute values of the shifts for the **F = 2** and **1** hyperfine states coincide to within the experimental error. The coincidence accuracy of calculated and experimental data depends, in particular, on the degree of polarization of atoms reached in the experiment and used in the calculation.

In (Dmitriev, 1997), the magnetic resonance frequency shifts of cesium atoms ^{133}Cs (**I$_A$ = 7/2**) were measured for two hyperfine states **F$_A$ = I$_A$ + 1/2 = 4** and **F$_A$ = I$_A$ − 1/2 = 3** in experiments on the optical orientation of atoms. An absorption chamber containing a mixture of ^{87}Rb (**I$_A$ = 3/2**) and ^{133}Cs in equal weight fractions and filled with the nitrogen buffer gas at a pressure of 100 Torr was placed in a thermostat. The concentration of alkali atoms in the absorption chamber was changed by varying the thermostat temperature from 20 to 80°C. The resonance radiation of the rubidium line was used for the optical orientation of Rb atoms. The degree of polarization of alkali atoms in experiments of this kind usually does not exceed few dozen percent. Then, the magnetic resonance was excited at the Zeeman frequency of cesium atoms for each of the hyperfine states, which resulted in a change of the absorption of the pumping light by rubidium atoms due to the spinexchange process between Cs and Rb. A change in the polarization of the pumping light from (+) to (−) allowed a doubled shift frequency of the magnetic resonance to be recorded. Figure 14 from (Kartoshkin, 2018) shows the dependences of the absolute values of doubled magnetic resonance frequency shift ($|2\delta f_\pm|$) on the temperature in the absorption chamber (± refers to F = 4 and 3, respectively).

The concentration of atoms of alkali metals was determined according to the tables given in (Nesmeyanov, 1963, 367). Since a mixture of alkali atoms was used in the experiment, to pass from the temperature of the absorption chamber to the concentration of alkali atoms in it, it is necessary to use Raoult's law for the pressure of the saturated vapor over the melt of a metal mixture. It is necessary to know the imaginary part of spin exchange complex cross section (1) in order to calculate the magnetic resonance frequency shifts determined by (9), (10). The temperature dependences of these quantities (q_{AA} and q_{AB}) were previously calculated for Cs–Cs and Rb–Cs pairs of atoms (Kartoshkin, 2018; Kartoshkin, 2016). Thus, it appears possible to calculate the magnetic resonance frequency shifts based on (9) and (10) using the calculated values of the cross sections. Figure 14 shows the results of this calculation for electron polarization of atom B ($P_z(B) = 0.3$ in this case, it is the Rb atom).

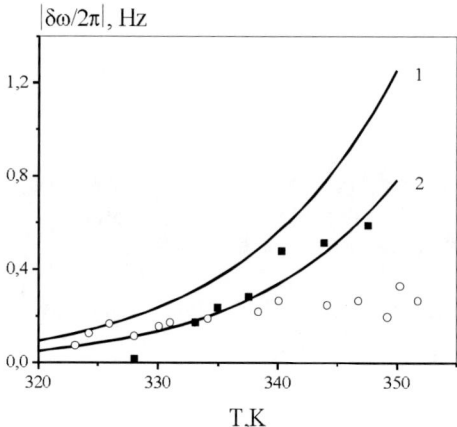

Figure 14. Dependences of the doubled magnetic resonance frequency shift of Cs atoms in a Cs–Rb mixture on temperature (Kartoshkin, 2018): (1) the calculated value of $2|\delta f|$ for the state **F = 3**, (2) the calculated value of $2|\delta f|$ for the state **F = 4**. Experimental values (symbols): (○) **F = 4**, (■) **F = 3** (Dmitriev and Dovator, 1997).

Conclusion

It can be seen from the results presented in this work that the shifts of the magnetic resonance frequency of alkali atoms in a mixture of these atoms depend significantly both on the values of the imaginary parts of the complex spin exchange cross sections and on their signs. In addition, since the expressions for frequency shifts include the average relative velocities of the motion of atoms, as well as the concentration of alkali atoms, the cross sections depend significantly on the temperature in the absorption chambers. In addition, the magnitude and sign of the shift depends on the magnitude of the nuclear spin of the atom for which the frequency shift is observed.

As follows from the obtained results, the temperature dependences of the dependence of the shift values pass through zero, which allows, by selecting the operating temperatures, to realize the situation when the magnetic resonance frequency shift vanishes, i.e., it becomes possible to avoid the negative influence of the spin exchange process on the system.

Disclaimer

None.

References

Ahmed E, Lyyra AM, Xie, F, Li, D, Chu, Y, Li, L, Ivanov, VS, Sovkov, VB, & Magnier, S. New Experimental Data on the K_2 $A^3\Sigma^+_u$ State Analyzed with the Multi-parameter Approach. *J. Mol. Spectrosc.* (2005) 234(1): 41-52.

Aleksandrov EB, Balabas MV, Vershovskii AK, and Pazgalev AS. A New Model of a Quantum Magnetometer: A Single-cell Cs-K Tandem Based on Four-quantum Resonance in ^{39}K Atoms. *Tech. Phys.* (2000) 45(7): 931-936.

Amiot C. The $X^1\Sigma_g^+$ Electronic State of K_2. *J. Mol. Spectrosc.* (1991) 146(2):370-382.

Amiot C, Verges J, and Fellows C. The Long-range Potential of the K_2 $X^1\Sigma^+_g$ Ground Electronic State up to 15 Å. *J. Chem. Phys.* (1995) 103(9):3350-3356.

Amiot C, and Dulieu O. The Cs_2 Ground Electronic State by Fourier Transform Spectroscopy: Dispersion Coefficients. *J. Chem. Phys.* (2002) 117(11):5155-5164.

Appelt S, Ben Amar Baranga A, Erickson, CJ, Romalis, MV, Young, AR and Happer W. "Theory of Spin-exchange Optical Pumping of ^3He and ^{129}Xe." *Phys. Rev. A* (1998) 58:1412-1439.

Dmitriev SP, Dovator NA. Observation of the Hyperfine Polarization and Alignmet in the Spin Exchange Cjllisions of the Polarized Cs and Rb. Atoms. *Opt. Spectrosc.* (1994) 77(5):712-713.

Dmitriev SP and Dovator NA. Experimental Determination of the Shift of the Zeeman Resonance Frequency of Cesium Atoms Induced by Spin-exchange Collisions with Rubidium Atoms. *Tech. Phys.* (1997) 42 (2):225-226.

Dmitriev SP and Dovator NA. Magnetic Resonance Frequency Shifts of Spin-Polarized Cesium Atoms in a Cs–Rb Mixture. *Tech. Phys.* (1997) 42(2):225-226.

Dmitriev SP and Dovator NA. Frequency Shift of the Magnetic Resonance in $5s^2S_{1/2}$ Rubidium Atoms Due to Spin-Exchange Collisions with Optically Oriented Cesium Atoms. *Tech. Phys.* (2007)52 (7):940-942.

Dmitriev SP, Dovator NA, and Kartoshkin VA. Spin Exchange upon Collision of Two Cesium Atoms in the Ground State. *Tech. Phys.* (2015) 60(6):826-829.

Ferber R, Klincare I, Nikolayeva O, Tamanis, M, Knöckel, H, Tiemann, E, & Pashov, A. The Ground Electronic State of KCs Studied by Fourier Transform Spectroscopy. *J. Chem. Phys.* (2008)128(24):244316.

Ferber R, Klincare I., Nikolayeva O, et al. $X^1\Sigma^+$ and $a^3\Sigma^+$ States of the Atom Pair K+Cs Studied by Fourier-transform Spectroscopy. *Phys. Rev. A* (2009) 80(6): 062501.

Gerdes A, Hobein M, Knöckel H, and Tiermann E. Ground State Potentials of the NaK Molecule. *Eur. Phys. J. D* (2008) 49 (1):67-73.

Happer W. Optical Pumping. *Rev. Mod. Phys.* (1972) 44(2):169-249.

Ivanov VS and Sovkov VB. Joint Analysis of the Attractive and Repulsive Regions of the Na_2 $a^3\Sigma^+_u$ State Potential: A New Empirical Potential Energy Curve. *J. Chem. Phys.* (2003) 118(18):8242-8247.

Jenc F. The Reduced Potential Curve Method for Diatomic Molecules and its applications. *Adv. Atom. Mol. Phys.* (1983) 119:265-305.

Jenc F, Brandt BA. Application of the Reduced-potential Curve Method for the Detection of Errors or Inaccuracies in the Analysis of Spectra and for the Construction of Internuclear Potentials of Diatomic Molecules: Alkali Diatomic Molecules. *Phys. Rev. A* (1989) 39(9): 4561-4582.

Kartoshkin VA. Complex Cross-sections for the Spin Exchange between Cs and Rb Alkali Atoms. *Opt. Spektrosc.* (1995) 79 (1): 26-31.

Kartoshkin VA. Magnetic Resonance Frequency Shifts upon Spin Exchange Collisions of Alkali Atoms. *Opt. Spectrosc.* (2010) 108(6):866–869.

Kartoshkin VA. Complex Cross Sections of Spin Exchange in Collisions of Sodium and Potassium Atoms in the Ground State. *Opt. Spectrosc.* (2010) 109(5):674-679.

Kartoshkin VA. Collisions of Alkali K and Rb Atoms in Ground State: Modification of Interaction Potentials. *Opt. Spectrosc.* (2011) 110(5):665-671.

Kartoshkin VA. Spin Exchange During Collisions of Potassium Atoms: Complex Cross Sections. *Opt. Spectrosc.* (2011) 111(12):881-884.

Kartoshkin VA. Spin-exchange Processes in a Single-chamber Cs-K Tandem Magnetometer. *Opt. Spectrosc.* (2012) 113(9): 235-239.

Kartoshkin VA. Complex Spin-Exchange Cross Sections in Collisions of Rubidium Isotopes in the Ground State. *Opt. Spectrosc.* (2015) 119(4):594-598.

Kartoshkin VA. Collisions of Alkali-Metal Atoms Cs and Rb in the Ground State. Spin Exchange Cross Sections. *Opt. Spectrosc.* (2016) 121(3): 327-330.

Kartoshkin VA. Magnetic-Resonance-Frequency Shifts in Spin-Exchange Collisions of Rb and Cs Alkali-Metal Atoms. *Opt. Spectrosc.* (2017) 123(3):335-337.

Kartoshkin VA. Magnetic Resonance Frequency Shifts of Spin-Polarized Cesium Atoms in a Cs–Rb Mixture. *Opt. Spectrosc.* (2018) 125(1):10-13.

Kartoshkin VA. Spin-Exchange Collisions in the K-Li System. *Opt. Spectr.*, (2019) 127(4):691-693.

Kartoshkin VA. Magnetic-Resonance Frequency Shifts in a Tandem Cs-K Magnetometer Induced by Spin Exchange. *Opt. Spectrosc.* (2020) 128(9):1355-1358.

Kartoshkin VA. Frequency Shifts of the Magnetic Resonance of Rb and K Atoms in the Rb-K Tandem Magnetometer. International Conference PhysicA.SPb/2020. *Journal of Physics: Conference Series* (2020) 1697:012146.

Kartoshkin VA. Collisions of Lithium Atoms in Ground State. Complex Spin-Exchange Cross Sections. *Opt. Spectrosc.* (2021) 130(6):641-644.

Kartoshkin VA. Magnetic Resonance Frequency Shift of Na Atoms during Collisions in a Mixture of Potassium and Sodium Atoms. *Tech. Phys.* (2021) 66(11):1221–1227.

Kartoshkin VA. Magnetic resonance frequency shifts of alkali atoms in the K-Li mixture. *Opt. Spectrosc.* (2022) 130(11):1634-1637. (in Russian).

Micalizio S, Godone A, Levi F, and Vanier J. Spin-exchange Frequency Shift in Alkali-metal-vapor Cell Frequency Standards. *Phys. Rev. A* (2006) 73(3):033414.

Mott NF, Massey HSW. 1965. *The Theory of Atomic Collisions*. Oxford: Clarendon Press.

Nesmeyanov AN. 1963. *Vapor Pressure of the Elements*. New York: Academic.

Okunevich AI. Spin-exchange Frequency Shifts for a Mixture of Alkali Atoms in an Inert Gas Atmosphere. *Opt. Spektrosc.* (1995)79 (5):718-728. (in Russian).

Pashov A, Docencko O, Tamanis M, Ferber, R, Knoeckel, H, Tiemann, E. Coupling of the $X^1\Sigma^+$ and $a^3\Sigma^+$ states of KRb. *Phys. Rev. A* (2007) 76(2): 022511.

Pomerantsev NM, Ryzhkov VM, and Skrotskii GV. 1972. *Physical Principles of Quantum Magnetometry*. Moscow: Nauka. (in Russian).

Vanier JV and Audoin C. 1989. *The Quantum Physics of Atomic Frequency Standards*. Bristol: Adam Hilger.

Wagshul ME and Chupp TE. Laser Optical Pumping of High-density Rb in Polarized ^{3}He Targets. *Phys. Rev. A* (1994) 49(5):3854-3869.

Xie F, Sovkov VB, Lyyra, AM, Li, D, Ingram, S, Bai, J, Ivanov, VS, Magnier, S and Li Li., Experimental Investigation of the Cs_2 $a\ ^3\Sigma_u^+$ Triplet Ground State: Multiparameter Morse Long Range Potential Analysis and Molecular Constants J. Chem. Phys. (2009)130(2): 051102.

Zemke WT and Stwalley WC. Analysis of Long Range Dispersion and Exchange Interactions between Two Na Atoms. *J. Chem. Phys.* (1994) 100(4):2661-2670.

Biographical Sketch

Victor Kartoshkin

Affiliation: Ioffe Institute, Division of Plasma Physics, Atomic Physics and Astrophysics, Lab. of Atomic Radiospectroscopy, St. Petersburg, Russian Federation

Education: In 1975 I graduated from the Radiophysics Faculty of the M. I. Kalinin Leningrad Polytechnic Institute (Now it is Peter the Great St. Petersburg Polytechnic University).

Business Address: 26 Politekhnicheskaya, St Petersburg 194021, Russian Federation

Research and Professional Experience: After graduating from the University in 1975 and to the present day, I am working at the Ioffe Institute. In 1980 I defended my Ph.D. thesis on the topic "Investigation of the process of metastability exchange in a mixture of helium isotopes He3-He4". In 1990 I defended my doctoral dissertation "Atom-Atomic and Atom-Molecular Interactions Involving Excited Polarized Atoms of Light Inert Gases". My scientific interests are focused on the study of collisional processes between spin-polarized atomic and molecular particles. Study of elastic and inelastic processes at the collision between such particles, transfer and conservation of spin polarization in such processes. The use of the results obtained in the creation of quantum electronics devices based on the principles of optical orientation of atoms. I have over 100 publicaations in peer-reviewed journals. The results of my research have been presented in more than 100 reports at various international conferences. I was the leader of a number of scientific projects in the field of applied and fundamental researches.

Professional Appointments: Chief Research Officer

Honors: International Society for Magnetic Resonance, member

Publications from the Last 3 Years:
1. Kartoshkin, VA. Angular Momentum Transfer at the Chemo-ionization and Spin Exchange Processes. Interaction between Spin Polarized Metastable Helium Atoms//2022, Mediterran. *J. Chem.*, v. 12 (2), 164-174.
2. Kartoshkin, VA. Сдвиги частоты магнитного резонанса щелочных атомов в смеси K-Li (Magnetic resonance frequency shifts of alkali atoms in the K-Li mixture.) 2022, *Оптика спектроск.*, т. 130 (11),1634-1637 doi: https://doi.org/10.21883/OS.2022.11.53767.3522-22 (in Russian).
3. Kartoshkin, VA. Collisions of Lithium Atoms in Ground State. Complex Spin-Exchange Cross Sections //2021, *Opt. Spectrosc.*, v. 129 (6), 641-644.
4. Kartoshkin, VA. Magnetic Resonance Frequency Shift of Na Atoms during Collisions in a Mixture of Potassium and Sodium Atoms//2021, *Tech. Phys.*, v. 66 (11), 1221-1227.
5. Kartoshkin, VA. Magnetic-Resonance Frequency Shifts in a Tandem Cs-K Magnetometer Induced by Spin Exchange//2020, *Opt. Spectrosc.*, v. 128 (9),1355-1358.

6. Kartoshkin, VA. Collisions with Participation of Polarized Cesium and Lithium Alkali Atoms//2020, *Opt. Spectrosc.*, v. 128 (4), 470-472.
7. Kartoshkin, VA. Frequency shifts of the magnetic resonance of Rb and K atoms in the K-Rb tandem magnetometer//*International Conference Physic A*. SPb/2020 *J. Phys.: Conf. Ser.*, v. 1697 (1), 01214.
8. Kartoshkin VA. Frequency shifts of the magnetic resonance of Rb and K atoms in the Rb-K tandem magnetometer. P. 241. . *International conference Physic A*. SPb/2019. 19-23 October 2020, Saint-Petersburg, Russia.
9. Kartoshkin VA. Frequency shift of the magnetic resonance at the spin exchange collisions between K and Li atoms. P. 3-30. *International conference Physic A*. SPb/2019. 18-22 October 2021, Saint-Petersburg, Russia.
10. Kartoshkin VA. Transfer of angular momentum at the interaction between spin polarized metastable helium atoms. P.107. *The Virtual 32nd International Conference on Photonic, Electronic and Atomic Collisions (ViCPEAC 2021)* from July 20-23, 2021. Canada.
11. Kartoshkin VA, Dmitriev, S, Dovator N. On line 27th International Symposium on Ion-Atom Collisions (27 ISIAC) July 14-16, 2021. Romania. Ionization of alkali atoms during the collisions with polarized metastable helium. *Redistribution of the spin polarization.* P. 30.
12. Kartoshkin VA. Collisions of spin polarized Cs atoms with alkali atoms. Spin exchange cross sections and magnetic resonance frequency shifts. P. 217. *International conference Physic A.* SPb/2022. 17-21 October 2022, Saint-Petersburg, Russia.

Chapter 2

Effects of Alkali Elements on Perovskite Halide Compounds

Takeo Oku[*]
Naoki Ueoka
Hayato Machiba
Atsushi Suzuki
Riku Okumura
and Ayu Enomoto
Department of Materials Science, The University of Shiga Prefecture, Hikone, Shiga, Japan

Abstract

Effects of doping with alkali elements such as Cs, Rb, K, or Na cations on $CH_3NH_3PbI_3$ perovskite photovoltaic cells were investigated and described. Lattice constants were slightly decreased and increased by K and Na doping, respectively, which indicated that Na atoms occupied interstitial sites in the perovskite crystal. Perovskite solutions with K and formamidinium (FA) iodides added were also used to fabricate perovskite solar cells. The addition of the K salts resulted in fine grained FA added perovskite crystals that were separated by very small gaps. These materials were used to fabricate solar cells that improved short circuit current densities, which resulted in an enhanced conversion efficiency, compared with standard materials. Effects of addition of alkali metal elements to Cu-doped $CH_3NH_3PbI_3$ photovoltaic devices

---

[*] Corresponding Author's Email: oku@mat.usp.ac.jp.

In: Alkali Metals
Editor: Wilbur M. Hulett
ISBN: 979-8-88697-706-6
© 2023 Nova Science Publishers, Inc.

were also investigated. The series resistance was decreased by simultaneous addition of CuBr$_2$ and RbI, which increased the external quantum efficiencies in the range of 300–500 nm, and the short-circuit current density. The stabilities can also be estimated by first-principle calculations, and the Gibbs energies were decreased by incorporation of alkali metal elements into the perovskite crystals. The copper d-orbital band was slightly above the valence-band maximum and functioned as an acceptor level for carrier generation. Excitation from iodine p-orbitals and copper d-orbitals to alkali metal s-orbitals could suppress carrier recombination and promote carrier transport.

Keywords: perovskite, alkali metals, structure, first-principles calculation, solar cell

Introduction

Perovskite solar cells are expected to be next-generation solar cells [1–6], and the photo-conversion efficiencies of the perovskite solar cells are currently reaching more than 25% [7, 8]. CH$_3$NH$_3$PbI$_3$ is a perovskite halide that exhibits semiconductor characteristics and has a broad absorption profile, which correlates with the high photoelectric conversion efficiencies of devices that contain CH$_3$NH$_3$PbI$_3$ with a tetragonal or cubic structure [9]. CH$_3$NH$_3$PbI$_3$ crystals that have been doped with metals, halogens and other compounds have been investigated, and the properties of solar cells are strongly dependent on the doped elements.

Introducing metal atoms such as tin (Sn) [10–14], germanium (Ge) [14–16], copper (Cu) [17–23], antimony (Sb) [24–28], zinc (Zn) [29–32], cobalt (Co) [32–34], arsenic (As) [35–37], manganese (Mn) [38,39], indium (In) [15], europium (Eu) [40, 41], samarium (Sm) [41], terbium (Tb) [41], or thallium (Tl) [15] at lead (Pb) sites has been performed. The optical absorption range of perovskite compounds has been extended by Sn or Tl doping [10, 15]. Doping sodium [42–44], potassium [44–47], rubidium [48–50], or cesium [51, 52] at the MA sites are expected to be effective for suppressing desorption of CH$_3$NH$_3$ sites in the MAPbI$_3$. Perovskite solar cells introduced with formamidinium (NH = CHNH$_3$, FA) [53–55], ethylammonium (CH$_3$CH$_2$NH$_3$, EA) [56–58] or guanidinium (C(NH$_2$)$_3$, GA) [59–62] at MA sites have also been developed and studied.

In the present article, effects of alkali elements to perovskite halide compounds were described. Partial-substitution structure models approximating perovskite compositions in actual devices were used to investigate the effects of alkali metals on the electronic structures. In addition, the electronic structures and thermodynamic stabilities were investigated by the first-principles molecular orbital and band calculations [63–66]. These studies will give us a guideline to develop perovskite halide compounds with alkali elements.

K-Doped $MA_{0.8}FA_{0.1}K_{0.1}PbI_3(Cl)$ Perovskite Solar Cells

The stability of a perovskite structure can be estimated using the tolerance factor (t) [9, 67–71]. If the t is between 0.90 and 1.0, the crystal has a cubic perovskite structure. If t is equal to 1, the perovskite crystal has an ideal cubic structure. The ionic radii of K^+, Pb^{2+}, I^-, and MA are 164, 119, 220 and 217 pm, respectively.

When KI, KBr, or KCl is added to $CH_3NH_3PbI_3$ at 10 at.%, the t are calculated to be 0.901, 0.902, or 0.903, respectively. Furthermore, $HC(NH_2)_2PbI_3$ perovskite compounds consisting of the formamidinium cation $HC(NH_2)_2^+$ (FA) have a more favorable band gap and charge transport properties when compared with $MAPbI_3$ [72, 73]. The ionic radius of FA is 253 pm [9], and doping of this large ionic radius causes the energy-gap of $CH_3NH_3PbI_3$-based perovskites to be reduced.

It was investigated how doping MA(FA)PbI$_3$(Cl) with K (i.e., KI, KBr, or KCl) affects the properties of photovoltaic devices. The doping was performed using an air-blowing method. Following the addition of KI, KBr, or KCl (10 at.%) to $MA_{0.8}FA_{0.1}K_{0.1}PbI_3(X)$, their respective t were calculated to be 0.907, 0.909, or 0.910 [46]. This indicated that the KBr or KCl-doped $MA_{0.8}FA_{0.1}K_{0.1}PbI_3(X)$ perovskite crystals had more stable cubic structure than that of the KI-doped perovskite. Doping effects of other alkali metals, such as Cs or Rb, to the MAPbI$_3$ have been investigated, and the photovoltaic properties were improved by the addition of Cs or Rb [74, 75]. It is also expected that K with a smaller ionic size would easily occupy the MA sites. K is also a cheap and abundant alkali metal, and so is a cheaper dopant than Rb or Cs. The addition of Cl^- is expected to increase the carrier diffusion length within the perovskite crystals [76, 77].

The fabrication and characterization of perovskite solar cells containing mesoporous TiO$_2$ as an electron transport layer, spiro-OMeTAD as a hole-transport layer and a KBr or KCl-doped CH$_3$NH$_3$PbI$_3$(Cl) perovskite as an active layer have been reported previously [46, 78]. An air-blowing method has been reported for permeating perovskite solutions throughout mesoporous TiO$_2$, which resulted in a highly (100)-oriented CH$_3$NH$_3$PbI$_3$(Cl) layer [79]. Using this method, we have synthesized K-doped perovskite crystals, denoted as MA$_{0.8}$FA$_{0.1}$K$_{0.1}$PbI$_3$(Cl, Br). The effects of doping the perovskite crystals with KBr or KCl on the properties of photovoltaic devices were investigated using a number of techniques, including light-induced current density–voltage (J-V) measurements, external quantum efficiency (EQE), X-ray diffraction (XRD), optical microscopy (OM), scanning electron microscopy (SEM), and energy dispersive X-ray spectroscopy (EDX).

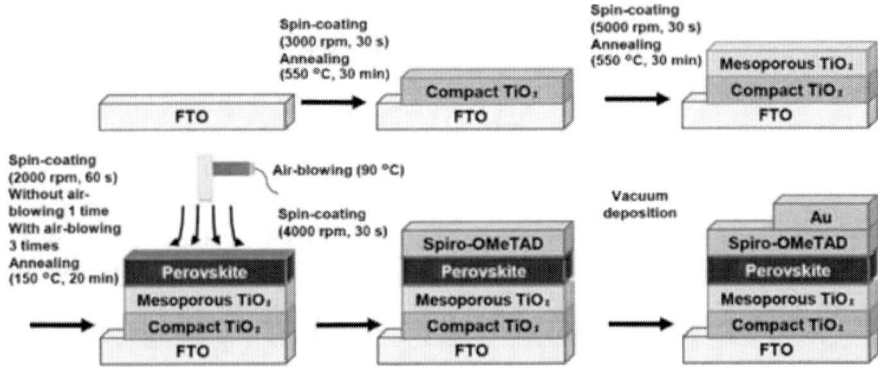

Figure 1. Schematic illustration showing the fabrication process of the perovskite solar cells.

A schematic illustration showing the fabrication processes used to fabricate photovoltaic cells is shown in Figure 1. All processes were performed in air [80, 81]. F-doped tin oxide (FTO) substrates were cleaned in an ultrasonic bath with acetone and ethanol, and dried under nitrogen gas. Subsequently, the FTO substrates were treated with an ultraviolet ozone cleaner (Asumi Giken ASM401N) for 15 min. TiO$_x$ precursor solutions (0.15 and 0.30 M) were prepared from titanium diisopropoxide bis(acetyl acetonate) (Sigma Aldrich) and 1-butanol (Wako Pure Chemical Industries). Both TiO$_x$ precursor solutions were spin-coated onto the FTO substrate at 3000 rpm for 30 s (Mikasa MSA100) and annealed at 125°C for 5 min, however, the 0.30M precursor was spin-coated twice to form a uniform layer. After that, the FTO substrate was sintered at 550°C for 30 min to form a compact TiO$_2$ layer. A mesoporous TiO$_2$ layer was spin-coated on top of the compact TiO$_2$ layer at 5000 rpm for 30 s using TiO$_2$ paste. The TiO$_2$ paste was prepared using TiO$_2$

powder (Aerosil P-25, 100 mg) and poly(ethylene glycol) (Nacalai Tesque PEG #20000, 10 mg) in distilled water (0.5 mL). This solution was mixed with acetylacetone (Wako Pure Chemical Industries, 10 μL) and the surfactant Triton X-100 (Sigma Aldrich, 5 μL) for 30 min and was then allowed to stand for ~24 h to suppress bubble formation within the solution. The FTO substrates with the TiO$_2$ were annealed at 550°C for 30 min to form the mesoporous TiO$_2$ layer.

The perovskite compounds were prepared by mixing solutions of CH$_3$NH$_3$I (Showa Chemical), HC(NH$_2$)$_2$I (Tokyo Chemical Industries), PbCl$_2$ (Sigma Aldrich), and KI (Wako Pure Chemical Industries), KBr (Wako Pure Chemical Industries), or KCl (Wako Pure Chemical Industries) at 70°C. All materials were dissolved in N,N-dimethylformamide (DMF; Sigma Aldrich). For comparison purposes, the standard precursor (MA$_{0.90}$FA$_{0.10}$PbI$_{3-x}$X$_x$) was prepared, and had molar concentrations of MAI, FAI, and PbCl$_2$ of 2.16, 0.24, and 0.8 M, respectively. MA$_{0.8}$FA$_{0.1}$K$_{0.1}$PbI$_3$(Cl) precursors, with concentrations of MAI, FAI, K compounds (KI, KBr, or KCl), and PbCl$_2$ of 1.92, 0.24, 0.24 and 0.8 M, respectively, were prepared. Solutions containing the perovskite precursors were spin-coated on the mesoporous TiO$_2$ layer at 2000 rpm for 60 s using simultaneous hot air-blowing to obtain highly oriented perovskite crystals [79]. The temperature of the FTO substrates during hot air-blowing was measured to be 90°C. Then, the cells were annealed at 150°C for 20 min in ambient air.

A hole transport layer (HTL) was formed by spin-coating at 4000 rpm for 30 s. The HTL precursor solution was prepared by adding 2,2',7,7'-tetrakis(N,N-di-p-methoxyphenylamine)-9,9'-spirobifluorene (Sigma Aldrich spiro-OMeTAD, 36.1 mg) to chlorobenzene (0.5 mL; Wako Pure Chemical Industries) and by stirring it for 24 h. A solution of lithium bis(trifluoromethylsulfonyl)imide (Li-TFSI, 260 mg; Tokyo Chemical Industry) in acetonitrile (0.5 mL; Nacalai Tesque) was also prepared by stirring it for 24 h. The former spiro-OMeTAD solution with 4-tert-butylpyridine (14.4 μL; Sigma Aldrich) was mixed with the latter Li-TFSI solution (8.8 μL) for 30 min at 70°C, and cooled to ambient temperature. Finally, a gold (Au) top-electrode was formed by vacuum evaporation.

The J-V characteristics of the photovoltaic devices that contained the K-doped compounds (MA$_{0.8}$FA$_{0.1}$K$_{0.1}$PbI$_3$(Cl)) are shown in Figure 2(a) [46]. The photovoltaic parameters of these devices are summarized in Table 1, where J_{SC} is the short-circuit current density, V_{OC} is the open-circuit voltage, FF is the fill factor, R_S is the series resistance, R_{sh} is the shunt resistance, η is the conversion efficiency, and η_{ave} is the average efficiency of four cells. The device that contained the standard material ($MA_{0.9}FA_{0.1}PbI_3(Cl)$) had a η of 12.94%, while the devices that contained the KI, KBr or KCl-doped perovskites had η 9.61, 8.96 and 14.15%, respectively. The J_{SC} and V_{OC} of the devices that contained the KCl-doped perovskite were higher than those of the standard device, which resulted in the improved conversion efficiency.

Table 1. Measured photovoltaic parameters of the present solar cells

Device	J_{SC} (mA cm^{-2})	V_{OC} (V)	FF	R_s (Ω cm^2)	R_{sh} (Ω cm^2)	η (%)	η_{ave} (%)
Standard	20.9	0.922	0.671	5.16	3562	12.94	12.31
+ 10% KI	19.6	0.915	0.536	9.07	1216	9.61	8.21
+ 10% KBr	20.7	0.833	0.519	7.94	238	8.96	7.54
+ 10% KCl	22.0	0.970	0.663	6.02	1794	14.15	13.26

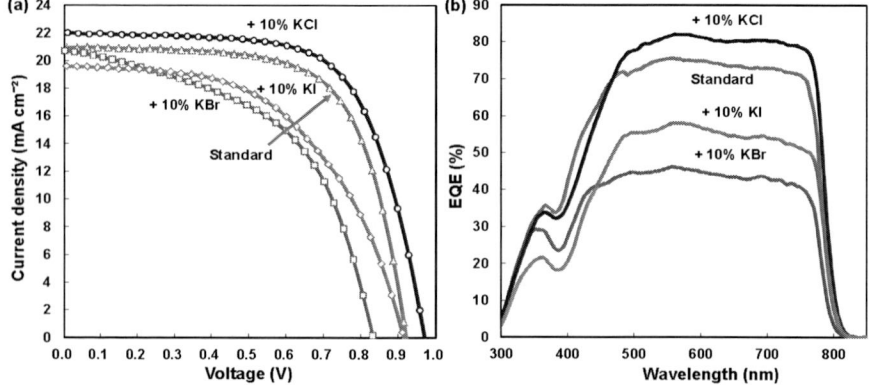

Figure 2. (a) *J-V* characteristics and (b) external quantum efficiencies of the perovskite solar cells that contained the standard and K-doped perovskites.

Table 2. Energy gaps, lattice constants and surface coverage of the perovskite crystals

Device	E_g (eV)	Lattice constant (Å)	Surface coverage (%)
Standard	1.54	6.278	95
+ 10% KI	1.54	6.292	87
+ 10% KBr	1.55	6.278	90
+ 10% KCl	1.53	6.295	96

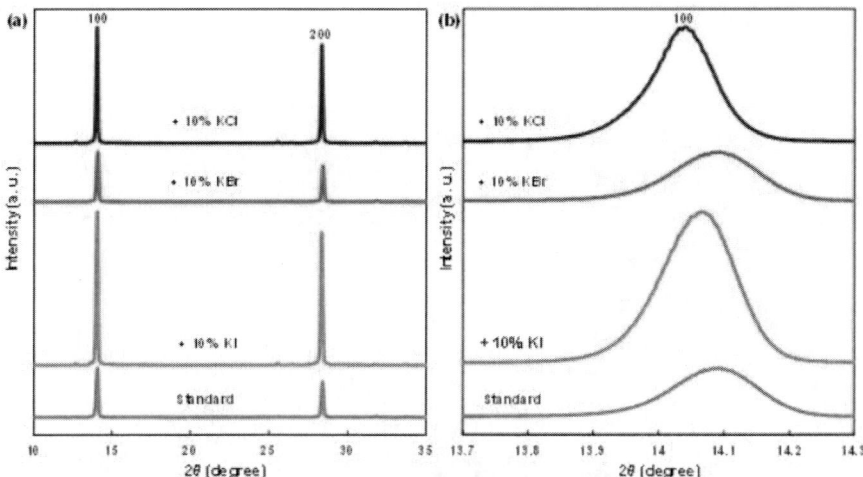

Figure 3. (a) X-ray diffraction patterns and (b) enlarged patterns of the perovskite solar cells that contained the standard and K-doped perovskites.

The EQE spectra of the photovoltaic devices are shown in Figure 2(b) and the estimated energy gaps (E_g) of the perovskite crystals, calculated by linear fitting using band gap calculator software (Enlitech), are summarized in Table 2. The energy gaps of the KI, KBr, or KCl-doped perovskite crystals were 1.54, 1.55 or 1.53 eV, respectively. Importantly, the energy gap of the KCl-doped perovskite crystals was narrower than that of the standard crystals, which led to an increase of the J_{SC} values. The EQE of the device that contained the KCl-doped perovskite was greater between 470 and 815 nm when compared with the standard device. This resulted in the increased J_{SC} that was observed. Conversely, the EQEs of the cells that contained the KBr or KI-doped perovskites were lower than the standard device over almost the entire the range examined.

XRD patterns of the standard $MA_{0.9}FA_{0.1}PbI_3(Cl)$ cells and the cells added with KI, KBr, or KCl are shown in Figure 3(a). An enlarged XRD patterns are also shown in Figure 3(b), which indicates a peak shift of 100 reflection to lower diffraction angles. The measured lattice constants (a) of the materials are summarized in Table 2. The devices with KI or KCl exhibited highly oriented (100) perovskite crystals. In addition, the (100) and (200) peak intensities increased significantly following the addition of KI or KCl. The device with KBr exhibited similar crystal orientation to the standard cell. The lattice constants of the KI and KCl-doped perovskites were higher than that of the standard material, whereas that of the KBr-doped perovskite did not change (Table 2). This will be discussed later in the schematic model of the lattice.

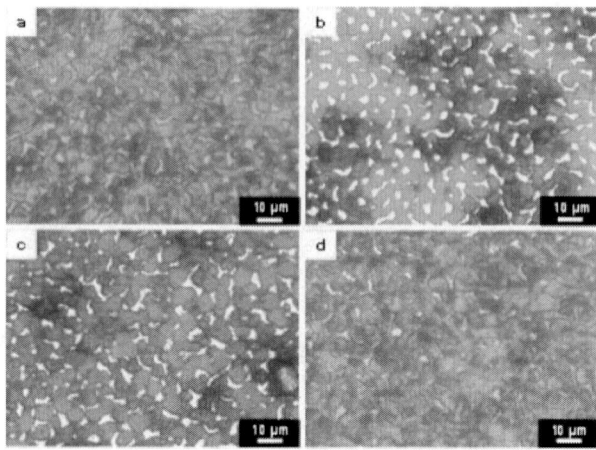

Figure 4. Optical microscope images of the perovskite solar cells that contained the standard and K-doped perovskites.

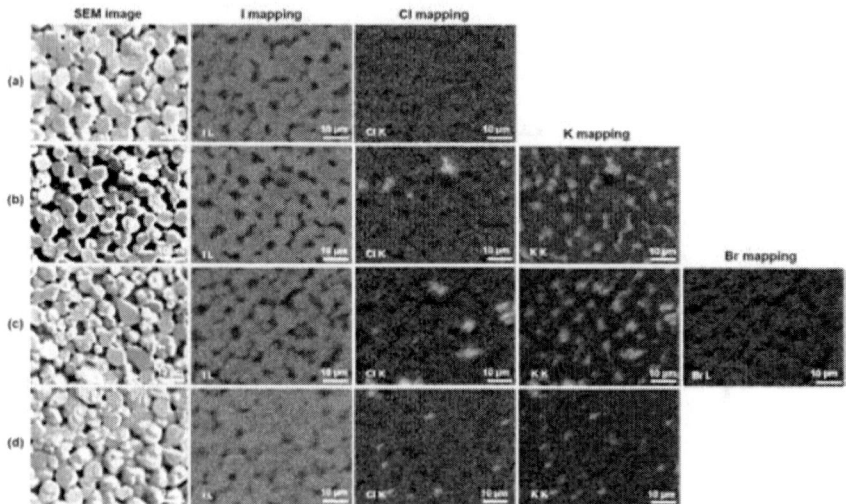

Figure 5. Scanning electron microscope images and energy dispersive X-ray spectrometry mapping of the (a) standard, (b) +10% KI, (c) +10% KBr, and (d) +10% KCl perovskite crystals.

Optical microscopy images of the devices are shown in Figure 4. The surface coverage of the perovskite layers was measured using ImageJ software [32], which resulted in surface coverage values of 95, 87, 90, and 96% for the standard, KI, KBr or KCl-doped perovskites, respectively (Table 2). The gaps between the grains of the KCl-doped perovskite crystals were very small when compared with those of the other materials, which resulted in the increased J_{SC} that was observed. On the other hand, the gaps between the grains of the KI- or KBr-doped perovskite crystals increased, which resulted in the decrease of FF values and conversion efficiencies.

SEM images and EDX surface mapping images (I, Cl, K, and Br) of the perovskite crystal surfaces are shown in Figure 5. The perovskite grains of the standard material ranged from 3 to 10 μm in size. Both I and Cl were distributed throughout the grains of each perovskite material. The KI-doped perovskite grains ranged from 2 to 5 μm in size and I, Cl and K were all distributed throughout the perovskite grains. The KBr-doped perovskite grains ranged from 2 to 7 μm in size and I, Cl, K and Br were distributed throughout the perovskite grains. The surface morphology of the KCl-doped perovskite crystals was significantly different than the other materials, with grains ranging from 3 to 8 μm in size that had covered the surface more effectively.

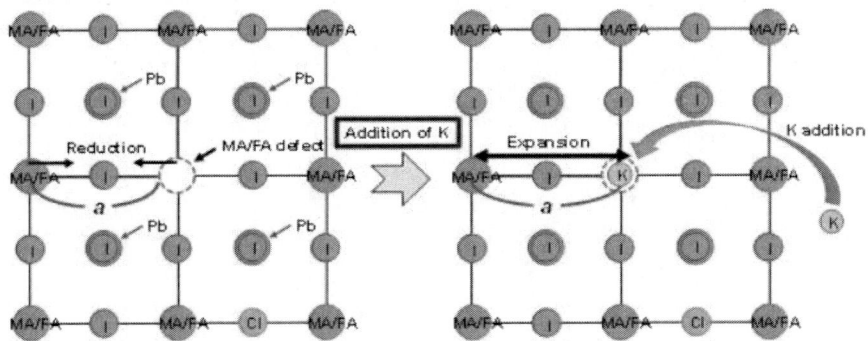

Figure 6. Schematic model showing the inclusion of potassium into the perovskite lattice and the resulting lattice expansion.

A schematic model showing the lattice structure of the K-doped perovskites is shown in Figure 6. The lattice constant of a perovskite single crystal [82] is larger than that of thin film crystals [79, 80]. When the perovskite crystals are formed as a thin film, lattice defects would be generated at MA/FA cation sites of the perovskite crystals. Since the ionic radius of K is significantly smaller than those of MA and FA, the resulting lattice constant of the K-doped crystals should be smaller than those of the standard material. However, the lattice constant of the KCl-doped crystals was larger than that of the standard material. It is plausible that this was caused by K introduction. Defects in the MA/FA cation sites within the perovskite crystals were produced during heat treatment, which resulted in the lattice shrinking. However, by adding the K salts, the K^+ ions were introduced into these defect positions, which would explain why the lattice constant of the perovskite crystals was observed to increase. Therefore, as the defects within the KCl-doped perovskite crystals had decreased, the resulting solar cells exhibited improved photoelectric conversion. The fact that the lattice constants of the KBr-doped perovskite crystals did not change when compared to the standard material indicated that the K^+ cations filled some of the defects. The Br^- ions, with a smaller ionic radius compared with I^-, also substituted the I^- ion sites. Further structure analysis using methods such as Reitveld refinement is required to confirm the occupancies and replacement of the K ions in the $CH_3NH_3PbI_3$ crystals [28].

The electronic structures of the perovskite crystals were calculated for the single-point with the experimental parameters obtained from XRD, through *ab-initio* quantum calculations based on the density functional theory using Gaussian 09 [66, 83]. The $MAPbI_3$ cubic structures were treated as a cluster model with supercells of $2 \times 2 \times 2$, as shown in Figure 7. A K element was introduced at the MA^+ sites at contents of 12.5%. Total density of states (DOS) of the present perovskite structures are shown in Figure 8. The electronic structure of $CH_3NH_3PbI_3$ mainly derives from the electronic state of the PbI_6 octahedron unit, and the level of the energy band is comprised of Pb 6s, 6p and I 5p orbits. The effect of MA void was found not to be so influential on the density of state of the perovskite, and the energy gap between highest occupied molecular orbital (HOMO) and lowest unoccupied molecular orbital (LUMO) decreased a little. It has also been reported that the charge-carrier trapping process has much faster timescale than the carrier recombination [84, 85]. On the other hand, the characteristic DOS of K 4s orbital in the range of 6–7 eV is observed as indicated by an arrow in Figure 8.

Increased generation of charge carriers by light absorption at short wavelengths could contribute to the improved EQE. Calculated Gibbs energies (G) of the standard $MAPbI_3$, MA defect and $MA(K)PbI_3$ models in Figure 7 were 946, 751, 743 kJ mol^{-1}, respectively. The values of G slightly decreased from 751 to 743 kJ mol^{-1}, and more stable perovskite structures were formed by the K introduction at the MA defect site. Therefore, the addition of the K element effectively stabilized the perovskite crystals.

Effects of Alkali Elements on Perovskite Halide Compounds 43

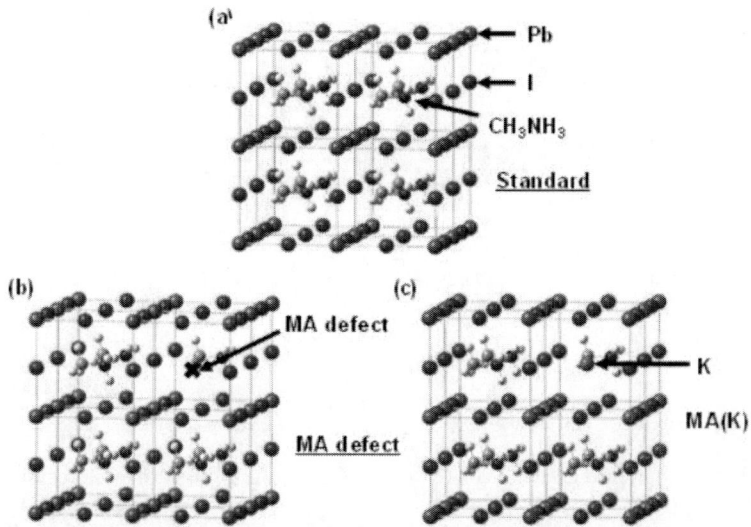

Figure 7. Structure models of (a) standard, (b) MA defect, and (c) K-doped perovskite structures.

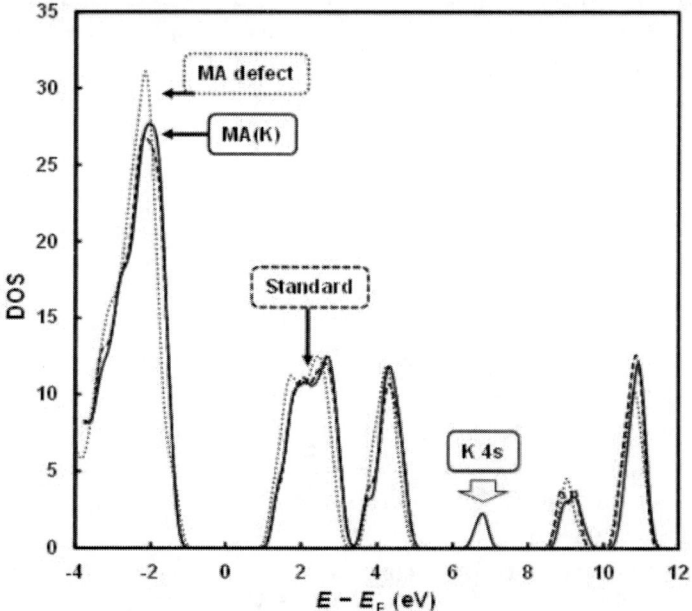

Figure 8. Total density of states of the perovskite structures in Figure 7.

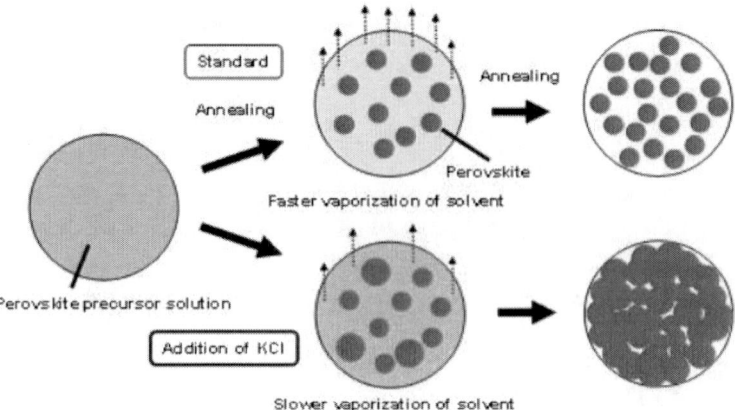

Figure 9. Schematic diagram showing the grain growth of the perovskite crystals.

A schematic showing a growth mechanism of the perovskite crystals is shown in Figure 9. In all systems, DMF was used as a solvent for the perovskite precursor solution. The boiling point of DMF is 153°C, and in the standard system evaporation occurred once the sample reached 150°C. Then, the growth of the perovskite particles would be too fast to sufficiently fill the grain gaps. Following the introduction of additives into the perovskite precursor solutions the boiling point of the DMF slightly increased, as can be determined using Eq. (1):

$$\Delta T = K_b \times m \quad (K_b = \frac{M \times R \times T_b}{\Delta H_b}) \tag{1},$$

where ΔT is the increase in the boiling point, K_b is the boiling point rise constant (K kg mol^{-1}), m the solute molar concentration (mol kg^{-1}), M is the solvent molar mass (kg mol^{-1}), R the gas constant (8.314 J K^{-1}), T_b is the boiling point of the pure solvent (K), and ΔH_b is the standard enthalpy of evaporation (J mol^{-1}). According to Eq. (1), ΔT is proportional to m, irrespective of the solute. Then a drop of the vapor pressure of the solvent occurs and the boiling point of the solution rises.

Thus, the addition of the K salts into the perovskite precursor solution increased the boiling point of the solutions, which resulted in slower vaporization of solvent and slower growth of the perovskite particles. SEM observations of the perovskite surfaces showed that the perovskite crystals did not grow as well in the KBr or KI-doped materials, but did in the KCl-doped material. This was likely caused by differences in the diffusion dependent reaction rates because of the different halogen anions. The diffusion coefficient can be described using Eq. (2)

$$D = \frac{k_B \times T}{6\pi \times v \times r_i} \tag{2},$$

where D is the diffusion coefficient, k_B is the Boltzmann constant (J K^{-1}), T is the temperature, v is the viscosity, and r_i is the ionic radius. In this work, only the ionic radius changed, and so the halogen would determine the value of the diffusion coefficient. In addition, the ionic radius of the A site (i.e., ABX$_3$ perovskite structure) was smaller than that of the standard system because of the small ionic radius of K, which would also act to decrease the diffusion coefficient. A diffusion-determined reaction can be represented by Eq. (3):

$$A + B \rightarrow AB \tag{3}.$$

Since the diffusion rate increased, the reaction rate increased. The rate constant (k_d) can be calculated using Eq. (4):

$$k_d = 4\pi N_A r_{AB}(D_A + D_B) \tag{4},$$

where N_A is the Avogadro's constant, r_{AB} is the distance between AB, and D_A and D_B are the diffusion coefficients of A and B, respectively. The rates of formation of the K-doped perovskite crystals increased in the following order: KCl > KBr > KI. As such, the perovskite crystals were more likely to be formed in this order, and the KCl-doped perovskite crystals would grow more efficiently because of the rise in boiling point and the faster diffusion rate.

Additive Effects of Alkali Metals on Cu-Doped CH$_3$NH$_3$PbI$_3$

Fabrication and characterization of Cu-added CH$_3$NH$_3$PbI$_{3-x}$Cl$_x$ photovoltaic devices with alkali metal elements were presented here [42]. First-principles calculations were performed to investigate the electronic structures. Elements with ionic radii smaller than MA$^+$ (2.17 Å) were added to the solution to compensate for the lattice distortion caused by Cu substitution at Pb sites. The effects of alkali metal iodides (i.e., NaI, KI, RbI, and CsI) and $CuBr_2$ addition were discussed here. It was expected that MA can be easily substituted by alkali metal elements in the perovskite crystal, and the alkali metal elements contribute to improved stability of the perovskite photovoltaic devices. The carrier transport mechanism was investigated by analyzing electronic structures and stabilities of the cells.

J–V characteristic under illumination recorded in the reverse scan and EQE spectra of FTO/TiO_2/perovskite/spiro-OMeTAD/Au photovoltaic devices are shown in Figure 10(a) and 10(b), respectively. Measured photovoltaic parameters of the present perovskite photovoltaic devices are summarized in Table 3. The standard cell provided a J_{SC} of 20.6 mA cm^{-2}, V_{OC} of 0.888 V, FF of 0.628, and η of 11.5%. The V_{OC} and η respectively increased from 0.888 V and 11.5% to 0.946 V and 12.6% through addition of $CuBr_2$. Although the series resistance (R_S) increased from 5.02 to 5.87 Ω cm^2 through $CuBr_2$ addition, R_S decreased to 4.77 Ω cm^2 by simultaneous addition of $CuBr_2$ and RbI. In addition, the shunt resistance (R_{Sh}) was increased from 2330 to 4664 Ω cm^2, and the leakage current was decreased by the $CuBr_2$ and RbI addition. As a result, η increased from 11.5% to 14.2% for the simultaneous addition of $CuBr_2$ and RbI. The FF also increased from 0.628 to 0.719 and the recombination between electrons and holes was suppressed by $CuBr_2$ and CsI addition.

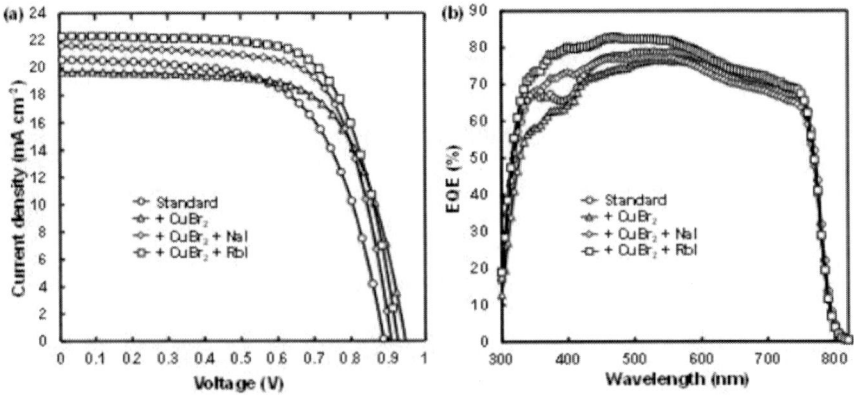

Figure 10. (a) J–V characteristics and (b) EQE of the present perovskite solar cells.

Table 3. Measured photovoltaic parameters of the present solar cells

Device	J_{SC} (mA cm^{-2})	V_{OC} (V)	FF	R_s (Ω cm^2)	R_{sh} (Ω cm^2)	η (%)	η_{ave} (%)
Standard	20.6	0.888	0628	5.02	2330	11.5	10.8
+ CuBr$_2$	19.7	0.946	0.674	5.87	2205	12.6	12.3
+ CuBr$_2$ + NaI	21.6	0.907	0.691	4.64	1068	13.6	12.8
+ CuBr$_2$ + KI	21.3	0.903	0.688	4.60	618	13.2	12.4
+ CuBr$_2$ + RbI	22.3	0.925	0.690	4.77	4664	14.2	13.8
+ CuBr$_2$ + CsI	20.7	0.897	0.719	4.05	1131	13.3	13.1

Table 4. Energy gaps, lattice constants and crystallite sizes of the perovskite crystals

Device	E_g (eV)	Lattice constant (Å)	Crystallite size (Å)
Standard	1.583	6.279	619
+ CuBr$_2$	1.585	6.282	617
+ CuBr$_2$ + NaI	1.587	6.274	630
+ CuBr$_2$ + KI	1.590	6.279	577
+ CuBr$_2$ + RbI	1.588	6.280	586
+ CuBr$_2$ + CsI	1.594	6.275	604

EQE values were increased by alkali metal elements addition in the range of 300–500 nm. The highest EQE of the device increased from 78.8% to 82.8% by RbI addition, indicating that the current density of the cells was increased by the RbI addition. The band gaps of the present perovskite photovoltaic devices estimated from EQE spectra are summarized in Table 4. The band gap was increased by addition of CuBr$_2$ and alkali metal elements. In particular, the band gap increased from 1.583 to 1.594 eV by CsI addition.

XRD patterns of the present photovoltaic devices are shown in Figure 11. The crystal structure of the standard cell was a cubic system, and the cells added with CuBr$_2$ and alkali metal elements also had a cubic system. Diffraction intensities of the 210 peak were the same in all devices. However, diffraction intensities of 100 increased through addition of NaI, RbI, and CsI. Hence, the 100 planes of the perovskite grains were preferentially oriented parallel to the FTO substrate [79]. Measured XRD parameters of the present perovskite photovoltaic devices are summarized in Table 4. The lattice constant was slightly increased by addition of CuBr$_2$, compared with the standard constant. Although the lattice constant might be expected to decrease owing to the smaller ionic radius of Cu and Br [86], the increase of the lattice distance suggested that lattice distortion charges became broadly distributed by replacing Pb with Cu. These results indicate that the charge carriers were generated more efficiently by the Cu substitution, as shown in Figure 10.

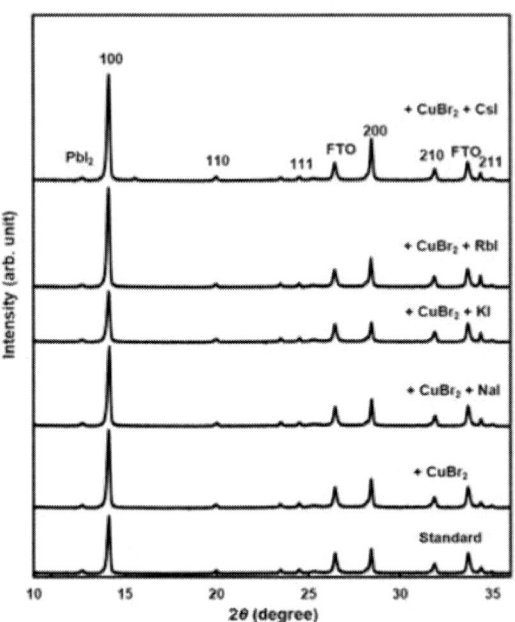

Figure 11. XRD patterns of the present perovskite photovoltaic devices.

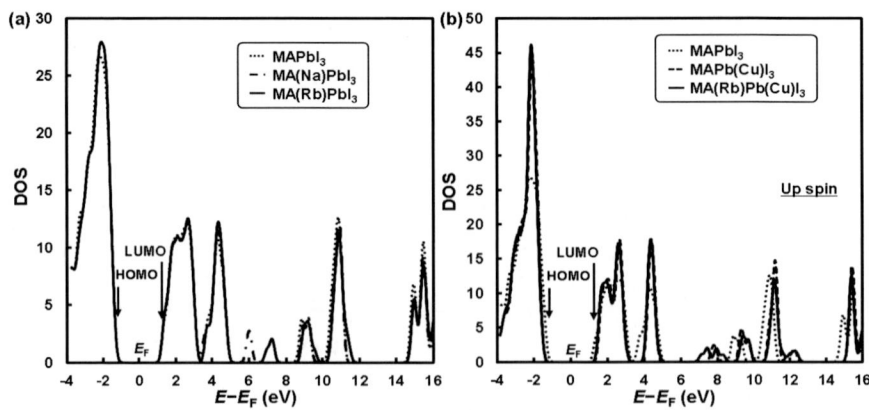

Figure 12. Total density of states of the present alkali-added perovskite structures (a) without and (b) with Cu.

In the case of Na substituted MA(Na)Pb(Cu)I$_3$, some differences were observed; however, the distribution of the electric charge was almost the same as that of MAPb(Cu)I$_3$. Because of the small ionic radius of Na, the interaction between Cu and Na is weak, resulting in minimal change of the electric charge distribution. Conversely, for MA(K)Pb(Cu)I$_3$, MA(Rb)Pb(Cu)I$_3$, and MA(Cs)Pb(Cu)I$_3$, we observed a change of the specific electric charge distribution of the HOMO level, owing to interactions of these alkali metal elements with Cu. The linear electric charge distribution would decrease the carrier transport resistance and increase the carrier diffusion length. Thus, the special electric charge distribution formed by CuBr$_2$ with NaI, KI, RbI and CsI contributes to the increase of J_{SC}. Total density of states with up-a spin and down-b spin of the present perovskite structures are shown in Figure 12(a) and 12(b), respectively. The electronic structure of CH$_3$NH$_3$PbI$_3$ mainly derives from the electronic state of the PbI6 octahedron unit, and the level of the energy band is comprised of Pb 6s, 6p and I 5p orbits. The energy level at the HOMO is a s anti-binding state between Pb 6s and I 5p orbits, and the LUMO is a mixed s or π anti-binding state between Pb 6p and I 5s or 5p orbits. Therefore, substitution of other elements at Pb sites influences the HOMO and LUMO levels. The DOS at the HOMO was greatly increased by Cu substitution but not influenced by the alkali metal substitution. Hence, the 3d orbital of Cu influences the energy level at HOMO and contributes to an increase in carrier generation. The characteristic increased DOS due to alkali metal elements is observed in the range of 5–10 eV, which would contribute the increased generation of charge carriers by light absorption at short wavelengths. Thus, the EQE improved in the range of 300–500 nm.

These results indicate that alkali metal elements make a contribution to the improved carrier transport properties. As shown in Figure 12(b), the energy level at the HOMO slightly approached the Fermi level and became sharper, which contributed to the increased hole mobility. Calculated energy levels at the HOMO and LUMO, and the energy gaps with a spin of the present perovskite structures are summarized in Table 5. The calculated E_g values are larger than those of experimental data in Table 4, and such differences have been previously reported for the cluster-model calculations. Effects of alkali metals on electronic structures at HOMO and LUMO are also shown in Figure 13.

The E_g increased from 2.70 to 3.23 eV by Cu substitution, which increased the V_{OC}. In addition, the energy gaps decreased in the order Cs$^+$ > Rb$^+$ > K$^+$ > Na$^+$. The hole transport properties between the perovskite and spiro-OMeTAD layers benefited from the shift of the DOS at the HOMO to the Fermi level and the decrease in the energy barrier. Therefore, the increase of the J_{SC} and V_{OC} are attributed to interactions of the Cu and alkali metals, as observed from their electronic structures and DOS.

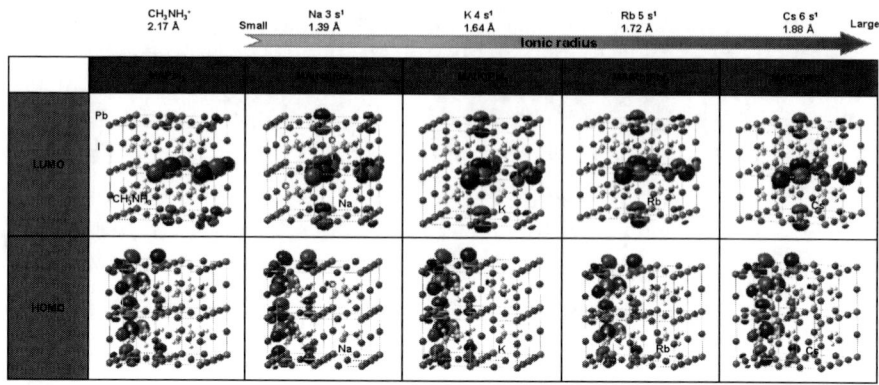

Figure 13. Electronic structures at HOMO and LUMO of the present perovskite structures with alkali metals.

Table 5. Calculated energy levels, energy gap and thermodynamic parameters of the present perovskite structures

Perovskite	LUMO (eV)	HOMO (eV)	E_g (eV)	G (kJ mol^{-1})	H (kJ mol^{-1})	S (kJ K^{-1} mol^{-1})
MAPbI$_3$	-14.9	-17.6	2.70	946	2326	4.63
MA(Na)PbI$_3$	-14.9	-17.6	2.70	728	2092	4.58
MA(K)PbI$_3$	-14.9	-17.6	2.70	741	2090	4.53
MA(Rb)PbI$_3$	-14.9	-17.6	2.70	721	2094	4.61
MA(Cs)PbI$_3$	-14.9	-17.6	2.70	722	2097	4.61
MAPb(Cu)I$_3$	-14.6	-17.8	3.23	958	2310	4.54
MA(Na)Pb(Cu)I$_3$	-14.6	-17.8	3.22	756	2074	4.42
MA(K)Pb(Cu)I$_3$	-14.6	-17.7	3.17	746	2076	4.46
MA(Rb)Pb(Cu)I$_3$	-14.6	-17.7	3.16	743	2078	4.48
MA(Cs)Pb(Cu)I$_3$	-14.6	-17.7	3.15	729	2084	4.55

Calculated thermodynamic parameters (G: Gibbs energy, H: enthalpy and S: entropy) of the present perovskite structure are summarized in Table 5. As a result of the calculation, the G, H, and S values of the MAPbI$_3$ perovskite structure were 946, 2326, and 4.63 kJ K^{-1} mol^{-1}, respectively. The value of G slightly increased from 946 to 958 kJ mol1 through Cu substitution, which increased the lattice constant of the perovskite with added CuBr$_2$, owing to the Jahn–Teller effect of Cu d electrons [87]. The Jahn–Teller effect is often observed for octahedron complexes with transition metals, where the repulsion between d orbital is relaxed by distortion of the arrangement of PbI_6 octahedra in the perovskite structure. Thus, addition of a large amount of transition metal distorts the perovskite structure. We expected that the lattice distortion of the perovskite crystal including Cu was relaxed by Na^+, K^+, Rb^+, and Cs^+ substitution. The value of G gradually decreased to 729 kJ mol^{-1}, and more stable perovskite structures were formed by alkali metal substitution in order of: $Cs^+ > Rb^+ > K^+ > Na^+$ substitution. Therefore, the substitution with alkali metal elements effectively stabilized the perovskite solar cells. However, further improvements to the solvent system are necessary because Cs compounds are poorly soluble in DMF.

The J_{SC}, V_{OC}, and h of the standard device were 17.4 mA cm^{-2}, 0.711 V and 6.7%, respectively, after 7 weeks. Conversely the device with added $CuBr_2$ and RbI provided the best J_{SC}, V_{OC}, and h of 21.8 mA cm^{-2}, 0.937 V and 14.0%, respectively after 7 weeks, which resulted in improved stability of the perovskite device. It was considered that the improved stability is associated with lattice defects. MA can be substituted easily by alkali metal elements in the perovskite crystal, and the alkali metal elements can contribute to the stability of perovskite photovoltaic devices. Moreover, the device with added $CuBr_2$ and CsI maintained the highest FF of more than 0.7 after 7 weeks. These results are consistent with our first-principles calculations.

Addition of Large Cations to Na-K-Cu-Doped CH$_3$NH$_3$PbI$_3$

Here, the effects of adding Cu, Na, and EA to the perovskite precursor solution were investigated with first-principles calculations [43, 88]. The addition of 2% Cu and 2% Na to MAPbI$_3$ improved the photoelectric conversion efficiency. Therefore, increased conversion properties and device durability by adding EA, together with 2-% Cu and 2-% Na, were examined. The calculations were performed to understand the doping effects in more detail, and to compare them with experimental results. Both total- and partial-substitution structure models approximating perovskite compositions in actual devices were used to investigate the effects.

The results in Table 6 indicated that co-addition of Cu and Na slightly increased the energy gap, the J_{SC}, V_{OC}, and FF, resulting in a higher conversion efficiency. The energy gap increased when I was replaced by Br, and the V_{OC} enhancement from the increased energy gap could be considered an effect of Br addition. The reason for the slight J_{SC} increase was attributed to increased carrier generation resulting from the acceptor-level Cu d-orbital band. The additional excitations from the Cu d-orbital to the Na s-orbital would also promote carrier generation. When K was added in addition to Cu, η increased compared with that of the standard device, which suggests that carrier separation is enhanced by co-addition of K and Cu. GA/K co-addition was not very effective in improving J_{SC}, which might be due to suppression of carrier transfer. Excess addition of Cu or K decreased the photovoltaic properties, which would be because of the poor crystallinity of the perovskites.

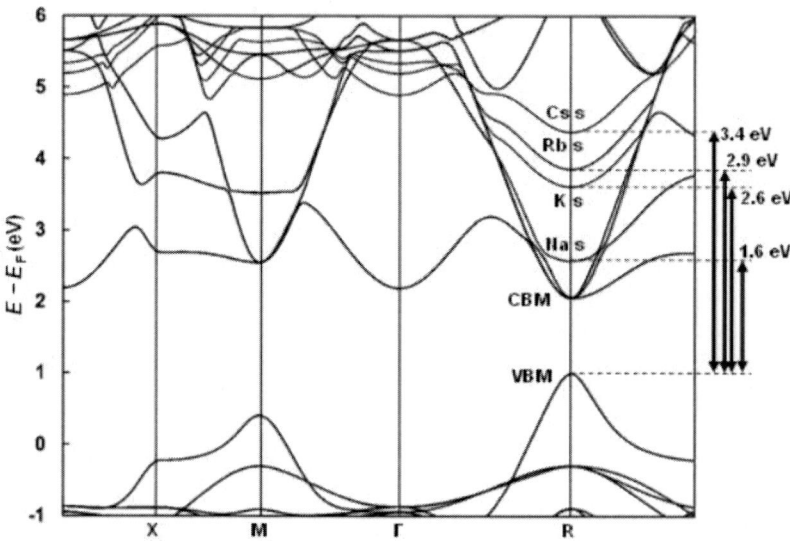

Figure 14. Changes of band structures for various alkali metals.

Table 6. Measured photovoltaic parameters of the present solar cells

Device	J_{SC} (mA cm^{-2})	V_{OC} (V)	FF	R_s (Ω cm^2)	R_{sh} (Ω cm^2)	η (%)	η_{ave} (%)
Standard	22.2	0.739	0.497	4.18	383	8.17	5.82
Cu 2% + Na 2%	21.7	0.841	0.578	5.93	594	10.5	8.36
Cu, Na 2% + EA 5%	21.3	0.843	0.585	5.12	558	10.5	8.67
Cu 2% + K 2%	21.4	0.837	0.590	6.26	513	10.6	8.99
Cu, K 2% + GA 10%	17.1	0.844	0.653	4.75	708	9.42	8.77

Table 7. Energy gaps, lattice constants and crystal orientation of the perovskite crystals

Device	E_g (eV)	Lattice constant (Å)	$I_{100}/I_{210}, I_{002}/I_{114}$
Standard	1.55	6.262(1)	8.7
Cu 2% + Na 2%	1.56	6.271(1)	2.6
Cu, Na 2% + EA 5%	1.57	6.261(1)	3.1
Cu 2% + K 2%	1.56	6.267(2)	6.2
Cu, K 2% + GA 10%	1.55	a = 8.885(4) c = 12.60(1)	9.9

Table 7 shows the measured structural parameters calculated from the XRD data. A diffraction peak corresponding to tetragonal (211) planes was observed at 23.5°. The peak intensity was slightly lower in the doped devices relative to that for MAPbI$_3$, suggesting that elimination of lattice distortion by the substitution of Pb with Cu reduced the peak intensity. Although the ionic radii of Cu, Na, and K are smaller than those of Pb and MA, their additions increased the lattice constant. The crystal lattice was expanded by the presence of Na in the interstitial position relative to that of MA. The lattice constants of devices doped with Cu, Na, and EA were the smallest, because of I substitution with Br, which had a smaller ionic radius. Crystallites of MAPbI$_3$ were larger than those of the doped devices. Previously, CuBr$_2$ addition to the perovskite precursor solution slowed the crystallization rate of the perovskite film, and decreased the crystallite size and diffraction-peak intensity. During annealing of the perovskite film, the color-changing rates with added Cu and Na were slower than those of MAPbI$_3$, indicating that the addition of CuBr$_2$ decreased the crystallization rate. The XRD patterns for the GA-added devices were indexed with a tetragonal structure. The structural distortion would be larger for the GA-added crystals, which would hinder carrier transfer and reduce the efficiency.

The alkali-metal energy bands formed above the conduction band minimum varied with the metal. These results are summarized in Figure 14 [43]. Substitution of MA with alkali metals significantly decreased the energy gap in the order Na, K, Rb, and Cs. Because the alkali-metal additions in the actual devices were a few mol%, the energy gaps were not expected to decrease as much as in the calculated results for the total-substitution models. The valence band maximum increased in the order Na, K, Rb, and Cs, which indicated that the addition of Rb or Cs increased the absorption of shorter-wavelength light in the visible region. The differences in energy levels in Figure 14 corresponded to energies that were higher than those of visible light.

However, because the Cu d-orbital bands were formed at the valence band maximum by adding Cu, the difference in energy levels between the alkali-metal bands and the Cu bands was approximately the same energy as that in the visible-light region.

Conclusion

Effects of doping with alkali elements such as Cs, Rb, K, or Na cations on $CH_3NH_3PbI_3$ perovskite photovoltaic cells were investigated. Lattice constants were slightly decreased and increased by K and Na doping, respectively. This indicated that Na atoms occupied interstitial sites in the perovskite crystal. Perovskite solutions with K and FA iodides added were also used to fabricate perovskite solar cells. The addition of the K salts resulted in fine grained FA added perovskite crystals that were separated by very small gaps. These materials were used to fabricate solar cells that improved short circuit current densities, which resulted in an enhanced conversion efficiency, compared with standard materials. The K occupied the MA defects within the perovskite crystals, which caused the lattice constant to increase. Additionally, the presence of K in the perovskite solutions promoted the growth of grains within the perovskite crystals. Effects of addition of alkali metal elements to Cu-doped $CH_3NH_3PbI_3$ photovoltaic devices were also investigated. The open-circuit voltage was increased by $CuBr_2$ addition to the $CH_3NH_3PbI_3$ precursor solution. The series resistance was decreased by simultaneous addition of $CuBr_2$ and RbI, which increased the external quantum efficiencies in the range of 300–500 nm, and the short-circuit current density. The stabilities can also be estimated by first-principle calculations, and several results were described. Calculations suggested that the Gibbs energies were decreased by incorporation of alkali metal elements into the perovskite crystals. The copper d-orbital band was slightly above the valence-band maximum and functioned as an acceptor level for carrier generation. Excitation from iodine p-orbitals and copper d-orbitals to alkali metal s-orbitals could suppress carrier recombination and promote carrier transport. Stabilities were improved by adding both Cu and Na, and simultaneous addition of Cu and K to the perovskite crystal suppressed decomposition of the crystal by desorption of MA, which led to improvement of the stability of the photovoltaic device.

Acknowledgments

This research was partly funded by Japan Society for the Promotion of Science as a Grant-in-Aid for Scientific Research (C) 21K04809.

References

[1] Jeong, M; Choi, I. W; Go, E. M; Cho, Y; Kim, M; Lee, B; Jeong, S; Jo, Y; Choi, H. W; Lee, J; Bae, J. H; Kwak, S. K; Kim, D. S; Yang, C. *Science*, 2020, 369, 1615−1620.
[2] Zhang, H; Ji, X; Yao, H; Fan, Q; Yu, B; Li, J. *Sol. Energy*, 2022, 233, 421–434.
[3] Lee, J. W; Tan, S; Il Seok, S; Yang, Y; Park, N. G. *Science*, 2022, 375, eabj1186.
[4] Li, N; Niu, X; Li, L; Wang, H; Huang, Z; Zhang, Y; Chen, Y; Zhang, X; Zhu, C; Zai, H; Bai, Y; Ma, S; Liu, H; Liu, X; Guo, Z; Liu, G; Fan, R; Chen, H; Wang, J; Lun, Y; Wang, X; Hong, J; Xie, H; Jakob, DS; Xu, XG; Chen, Q; Zhou, H. *Science*, 2021, 373, 561–567.
[5] Lin, R; Xu, J; Wei, M; Wang, Y; Qin, Z; Liu, Z; Wu, J; Xiao, K; Chen, B; Park, S. M; Chen, G; Atapattu, H. R; Graham, K. R; Zhu, J; Li, L; Zhang, C; Sargent, E. H; Tan, H; *Nature*, 2022, 603, 73–78.
[6] Sharma, R; Sharma, A; Agarwal, S; Dhaka, M. S. *Sol. Energy*, 2022, 244, 516−535.
[7] Yoo, J. J; Seo, G; Chua, M. R; Park, T. G; Lu, Y; Rotermund, F; Kim, Y. K; Moon, C. S; Jeon, N. J; Correa-Baena, J. P; Bulović, V; Shin, S. S; Bawendi, M. G; Seo, J. *Nature*, 2021, 590, 587−593.
[8] Zhao, Y; Ma, F; Qu, Z; Yu, S; Shen, T; Deng, H. X; Chu, X; Peng, X; Yuan, Y; Zhang, X; You, J. *Science*, 2022, 377, 531−534.
[9] Oku, T. *Rev. Adv. Mater. Sci.*, 2020, 59, 264−305.
[10] Hao, F; Stoumpos, C. C; Cao, D. H; Chang, R. P. H; Kanatzidis, M. G. *Nat. Photonics*, 2014, 8, 489−494.
[11] Liu, H; Zhang, Z; Zuo, W; Roy, Ra; Li, Meng; Byranvand, M. M; Saliba, M. *Adv. Energy Mater.*, 2022, 2202209.
[12] Zhang, M; Zhang, Z; Cao, H; Zhang, T; Yu, H; Du, J; Shen, Y; Zhang, XL; Zhu, J; Chen, P; Wang, M. *Mater. Today Energy*, 2022, 23, 100891.
[13] Asakawa, Y; Oku, T; Kido, M; Suzuki, A; Okumura, R; Okita, M; Fukunishi, S; Tachikawa, T; Hasegawa, T. *Technologies*, 2022, 10, 112.
[14] Krishnamoorthy, T; Ding, H; Yan, C; Leong, W. L; Baikie, T; Zhang, Z; Sherburne, M; Li, S; Asta, M; Mathews, N; Mhaisalkarac, S. G. *J. Mater. Chem. A*, 2015, 3, 23829−23832.
[15] Ohishi, Y; Oku, T; Suzuki, A. *AIP Conf. Proc.*, 2016, 1709, 020020-1-8.
[16] Tanaka, H; Ohishi, Y; Oku, T. *AIP Conf. Proc.*, 2018, 1929, 020007-1− 7.
[17] Shirahata, Y; Oku, T. *Phys. Stat. Solidi A*, 2017, 214, 1700268-1-6.
[18] Shirahata, Y; Oku, T. *Mater. Res. Express*, 2018, 5, 055504-1–11.
[19] Tanaka, H; Ohishi, Y; Oku, T. *Jpn. J. Appl. Phys.*, 2018, 57, 08RE10-1− 5.
[20] Tanaka, H; Oku, T; Ueoka, N. *Synth. Metals*, 2018, 244, 128–133.

[21] Oku, T; Ohishi, Y; Tanaka, H. *AIP Conf. Proc.*, 2018, 1929, 020010-1– 8.
[22] Ueoka, N; Oku, T. *ACS Appl. Energy Mater.*, 2020, 3, 7272–7283.
[23] Ueoka, N; Oku, T; Suzuki, A. *AIP Adv.*, 2020, 10, 125023.
[24] Oku, T; Ohishi, Y; Suzuki, A. *Chem. Lett.*, 2016, 45, 134–136.
[25] Oku, T; Ohishi, Y; Suzuki, A; Miyazawa, Y. *Metals*, 2016, 6, 147-1–13.
[26] Zhang, J; Shang, MH; Wang, P; Huang, X; Xu, J; Hu, Z; Zhu, Y; Han, L. *ACS Energy Lett.*, 2016, 1, 535–541.
[27] Oku, T; Ohishi, Y; Suzuki, A. *AIP Conf. Proc.*, 2017, 1807, 020007-1-5.
[28] Ando, Y; Oku, T; Ohishi, Y. *Jpn. J. Appl. Phys.*, 2018, 57, 02CE02-1–5.
[29] Tanaka, H; Ohishi, Y; Oku, T. *AIP Conf. Proc.*, 2018, 1929, 020007-1– 7.
[30] Zhao, W; Yang, D; Yang, Z; Liu, S. *Mater. Today Energy*, 2017, 5, 205– 213.
[31] Zhang, X; Yin, J; Nie, Z; Zhang, Q; Sui, N; Chen, B; Zhang, Y; Qu, K; Zhao, J; Zhou, H. *RSC Advances*, 2017, 7, 37419–37425.
[32] Long, Y; Wang, C; Liu, X; Wang, J; Fu, S; Zhang, J; Hu, Z; Zhu, Y. *J. Mater. Chem. C*, 2021, 9, 2145–2155.
[33] Klug, M. T; Osherov, A.; Haghighirad, A. A; Stranks, S. D; Brown, P. R; Bai, S.; Wang, J. T. W; Dang, X; Bulovi, V; Snaith, H. J; Belcher, A. M. *Energy Environ. Sci.*, 2017, 10, 236.
[34] Xu, W; Zheng, L; Zhang, X; Cao, Y; Meng, T; Wu, D; Liu, L; Hu, W; Gong, X. *Adv. Energy Mater.*, 2018, 8, 1703178.
[35] Suzuki, A; Oe, M; Oku, T. *J. Electron. Mater.*, 2021, 50, 1980–1995.
[36] Hamatani, T; Shirahata, Y; Ohishi, Y; Fukaya, M; Oku, T. *Adv. Mater. Phys. Chem.*, 2017, 7, 1–10.
[37] Hamatani, T; Shirahata, Y; Ohishi, Y; Fukaya, M; Oku, T. *AIP Conf. Proc.*, 2017, 1807, 020012-1–9.
[38] Hamatani, T; Oku, T. *AIP Conf. Proc.*, 2018, 1929, 020018-1-8.
[39] Zhang, H; Wang, H; Williams, S. T; Xiong, D; Zhang, W; Chueh, C. C; Chen, W; Jen, A. K. Y. *Adv. Mater.*, 2017, 1606608-1–8.
[40] Taguchi, M; Suzuki, A; Tanaka, H; Oku, T. *AIP Conf. Proc.*, 2018, 1929, 020012-1–8.
[41] Suzuki, A; Oku, T. *Mater. Adv.*, 2021, 2, 2609–2616.
[42] Suzuki, A; Kishimoto, K; Oku, T; Okita, M; Fukunishi, S; Tachikawa, T. *Synth. Met.* 2022, 287, 117092.
[43] Ueoka, N; Oku, T; Suzuki, A. *RSC Adv.*, 2019, 9, 24231–24240.
[44] Okumura, R; Oku, T; Suzuki, A; Okita, M; Fukunishi, S; Tachikawa, T; Hasegawa, T. *Appl. Sci.*, 2022, 12, 1710.
[45] Suzuki, A; Kitagawa, K; Oku, T; Okita, M; Fukunishi, S; Tachikawa, T. *Electron. Mater. Lett.*, 2021, 1–11.
[46] Tang, Z; Bessho, T; Awai, F; Kinoshita, T; Maitani, M. M; Jono, R; Murakami, T. N; Wang, H; Kubo, T; Uchida, S; Segawa, H. *Sci. Rep.*, 2017, 7, 12183-1–6.
[47] Machiba, H; Oku, T; Kishimoto, T; Ueoka, N; Suzuki, A. *Chem. Phys. Lett.* 2019, 730, 117–123.
[48] Kandori, S; Oku, T; Nishi, K; Kishimoto, T; Ueoka, N; Suzuki, A. *J. Ceram. Soc. Jpn.*, 2020, 128, 805–811.

[49] Saliba, M; Matsui, T; Domanski, Konrad; Seo, J. Y; Ummadisingu, Amita; Zakeeruddin, S. M; Correa-Baena, J. P; Tress, W. R; Abate, A; Hagfeldt, Anders; Grätzel, M. *Science*, 2016, 354, 206–209.
[50] Turren-Cruz, S. H; Saliba, M; Mayer, M. T; Juárez-Santiesteban, H; Mathew, X; Nienhaus, L; Tress, W; Erodici, M. P; Sher, M. J; Bawendi, M. G; Grätzel, M; Abate, A; Hagfeldt, A; Correa-Baena, J. P. *Energy Environ. Sci.*, 2018, 11, 78–86.
[51] Takada, K; Oku, T; Suzuki, A; Okita, M.; Fukunishi, S.; Tachikawa, T.; Hasegawa, T. *Chem. Proc.* 2022, 9, 14.
[52] Ueoka, N; Oku, T; Ohishi, Y; Tanaka, H; Suzuki, A; Sakamoto, H; Yamada, M; Minami, S; Tsukada, S. *AIP Conf. Proc.*, 2018, 1929, 020026-1–8.
[53] Ueoka, N; Oku, T; Suzuki, A; Sakamoto, H; Yamada, M; Minami, S; Miyauchi, S. *Jpn. J. Appl. Phys.*, 2018, 57, 02CE03-1–7.
[54] Bi, D; Tress, W; Dar, M. I; Gao, P; Luo, J; Renevier, C; Schenk, K; Abate, A; Giordano, F; Baena, J. P. C; Decoppet, J. D; Zakeeruddin, S. M; Nazeeruddin, M. K; Grätzel, M; Hagfeldt, A. *Sci. Adv.*, 2016, 2, e1501170-1–7.
[55] Suzuki, A; Kato, M; Ueoka, N; Oku, T. *J. Electron. Mater.*, 2019, 48, 3900–3907.
[56] Suzuki, A; Taguchi, M; Oku, T; Okita, M; Minami, S; Fukunishi, S; Tachikawa, T. *J. Mater. Sci. Mater. Electron.*, 2021, 32, 26449–26464.
[57] Peng, W; Miao, X; Adinol, V; Alarousu, E; Tall, O. E; Emwas, A. H; Zhao, C; Walters, G; Liu, J; Ouellette, O; J. Pan, Murali, B; Sargent, E. H; Mohammed, O. F; Bakr, O. M. *Angew. Chem., Int. Ed.*, 2016, 55, 10686–10690.
[58] Liu, D; Li, Q; Wu, K. *RSC Adv.*, 2019, 9, 7356.
[59] Nishi, K; Oku, T; Kishimoto, T; Ueoka, N; Suzuki, A. *Coatings*, 2020, 10, 410.
[60] Jodlowski, A. D; Roldán-Carmona, C; Grancini, G; Salado, M; Ralaiarisoa, M; Ahmad, S; Koch, N; Camacho, L; de Miguel, G. Nazeeruddin, MK. *Nat. Energy*, 2017, 2, 972–979.
[61] Kishimoto, T; Suzuki, A; Ueoka, N; Oku, T. *J. Ceram. Soc. Jpn.*, 2019, 127, 491–497.
[62] Ono, I; Oku, T; Suzuki, A; Asakawa, Y; Terada, S; Okita, M; Fukunishi, S; Tachikawa, T. *Jpn. J. Appl. Phys.*, 2021, 61, SB1024.
[63] Kishimoto, T; Oku, T; Suzuki, A; Ueoka, N. *Phys. Stat. Solidi A*, 2021, 218, 2100396.
[64] Körbel, S; Marques, M. A. L; Botti, S. *J. Mater. Chem. C*, 2016, 4, 3157- 3167.
[65] Suzuki, A; Oku, T. *Jpn. J. Appl. Phys.*, 2018, 57, 02CE04-1–7.
[66] Suzuki, A; Oku, T. *Heliyon*, 2018, 4, e00755-1–22.
[67] Suzuki, A; Oku, T. *Appl. Surf. Sci.*, 2019, 483, 912–921.
[68] Green, M. A; Ho-Baillie, A; Snaith, H. J. *Nat. Photonics*, 2014, 8, 506–514.
[69] Hoefler, S. F; Trimmel, G; Rath, T. *Monatsh. Chem.*, 2017, 148, 795–826.
[70] Xu, F; Zhang, T; Li, G; Zhao, Y. *J. Mater. Chem. A*, 2017, 5, 11450–11461.
[71] Sampson, M. D; Park, J. S; Schaller, R. D; Chan, M. K. Y; Martinson, A. B. F. *J. Mater. Chem. A*, 2017, 5, 3578–3588.
[72] Tanaka, H; Oku, T; Ueoka, N. *Jpn. J. Appl. Phys.*, 2018, 57, 08RE12.
[73] Amat, A; Mosconi, E; Ronca, E; Quarti, C; Umari, P; Nazeeruddin, M. K; Grätzel, M; Angelis, FD. *Nano Lett.*, 2014, 14, 3608–3616.

[74] Slimi, B; Mollar, M; Assaker, I. B; Kriaa, I; Chtourou, R; Mari, B. *Energy Procedia*, 2016, 102, 87–95.
[75] Hu, Y; Hutter, E. M; Rieder, P; Grill, I; Hanisch, J; Aygüler, M. F; Hufnagel, A. G; Handloser, M; Bein, T; Hartschuh, A; Tvingstedt, K; Dyakonov, V; Baumann, A; Savenije, T. J; Petrus, M. L; Docampo, P. *Adv. Energy Mater.*, 2018, 8, 1703057.
[76] Zhang, M; Yun, J. S; Ma, Q; Zheng, J; Lau, C. F. J; Deng, X; Kim, J; Kim, D; Seidel, J; Green, M. A; Huang, S; Ho-Baillie, A. W. Y. *ACS Energy Lett.*, 2017, 2, 438–444.
[77] Stranks, S. D; Eperon, G. E; Grancini, G; Menelaou, C; Alcocer, M. J. P; Leijtens, T; Herz, L. M; Petrozza, A; Snaith, H. J. *Science*, 2013, 342, 341–344.
[78] Dong, Q; Fang, Y; Shao, Y; Mulligan, P; Qiu, J; Cao, L; Huang, J. *Science*, 2015, 347, 967–970.
[79] Machiba, H; Oku, T; Suzuki, A. *AIP Conf. Proc.* 2019, 2067, 020009.
[80] Oku, T; Ohishi, Y; Ueoka, N. *RSC Adv.*, 2018, 8, 10389–10395.
[81] Oku, T; Zushi, M; Imanishi, Y; Suzuki, A; Suzuki, K. *Appl. Phys. Express*, 2014, 7, 121601.
[82] Oku, T; Ohishi, Y. *J. Ceram. Soc. Jpn.*, 2018, 126, 56–60.
[83] Ren, Y; Oswald, I. W. H; Wang, X; McCandless, G. T; Chan, J. Y. *Crystal Growth Design*, 2016, 16, 2945–2951.
[84] Suzuki, A; Miyamoto, Y; Oku. T, *J. Mater. Sci.*, 2020, 55, 9728–9738.
[85] Yang, B; Chen, J; Hong, F; Mao, X; Zheng, K; Yang, S; Li, Y; Pullerits, T; Deng, W; Han, K. *Angew. Chem. Int. Ed.*, 2017, 56, 12471–12475.
[86] Yang, B; Mao, X; Hong, F; Meng, W; Tang, Y; Xia, X; Yang, S; Deng, W; Han, K. *J. Am. Chem. Soc.*, 2018, 140, 17001–17006.
[87] Jahandar, M; Heo, J. H; Song, C. E; Kong, K. J; Shin, W. S; Lee, J. C; Im, SH; Moon, S. J. *Nano Energy*, 2016, 27, 330–339.
[88] Cortecchia, D; Dewi, H. A; Yin, J; Bruno, A; Chen, S; Baikie, T; Boix, P. P; Grätzel, M; Mhaisalkar, S; Soci, C; Mathews, N. *Inorg. Chem.*, 2016, 55, 1044–1052.
[89] Enomoto, A; Suzuki, A; Oku, T; Okita, M; Fukunishi, S; Tachikawa, T; and T. Hasegawa. *J. Electron. Mater.*, 2022, 51, 4317–4328.

Chapter 3

The Interaction between Spin Polarized Cesium Atoms and Alkali Atoms: Spin Exchange Collisions

Victor A. Kartoshkin[*]
Division of Plasma Physics, Atomic Physics and Astrophysics, Ioffe Institute, St. Petersburg, Russia

Abstract

The study of interactions involving spin-polarized atoms began quite a long time ago, but in recent years, interest in these interactions has again become topical. This is due to both the development of experimental technique, in particular, semiconductor lasers, including vertical cavity surface-emitting lasers (VCSEL), which, with their miniature size, allow to achieve a high degree of polarization of alkali atoms, in particular, cesium. This kind of miniaturization made it possible to consider the possibilities of designing devices, the principle of operation of which is based on the optical orientation of atoms. Such devices include quantum gyroscopes, magnetoencephalographs, and quantum magnetometers.

The method of optical orientation of atoms makes it possible to obtain spin-polarized optical particles. The presence of polarized atoms in the working chambers of the above devices leads to the fact that collisions between atoms of the working mixtures make it possible to obtain spin polarized atoms that did not interact with the pumping light (indirect optical orientation of atoms). These collisions lead not only to the transfer of spin polarization to the collision partners, but also to the broadening of the magnetic resonance lines and to a shift of the frequency of the magnetic resonance lines. The paper considers the interaction of

---

[*] Corresponding Author's Email: victor.kart@mail.ioffe.ru.

In: Alkali Metals
Editor: Wilbur M. Hulett
ISBN: 979-8-88697-706-6
© 2023 Nova Science Publishers, Inc.

spin polarized cesium atoms with alkali atoms Cs, Rb, K, and Li in the ground state.

Collisions of alkali-metal atoms in the ground state with the electron spin S = 1/2 are accompanied by exchange of electron coordinates between the colliding particles, which leads to the polarization transfer between them (i.e., to the well-known phenomenon of spin exchange). In addition, along with the polarization transfer from one partner to another, the magnetic resonance lines of colliding atoms are broadened and shifted in spin-exchange collisions. The last two processes depend, in particular, on the complex spin-exchange cross sections. The real part of the cross section determines the so-called "spin exchange cross section," which is responsible for broadening of magnetic resonance lines, while the imaginary part - the shift cross section - governs the frequency shift of magnetic resonance. The spin exchange broadening of a magnetic resonance line affects the precision of such quantum electronics devices as quantum frequency standards and magnetometers.

To describe the spin exchange process, we have to know complex spin exchange cross sections. Complex spin-exchange cross sections are calculated based on the data on the singlet ($X^1\Sigma^+$) and triplet ($a^3\Sigma^+$) potentials describing the interaction Cs alkali-metal atoms in the ground state with other alkaline atoms. The cross-section values allow to calculate the processes of polarization transfer and the relaxation times, as well as the magnetic resonance frequency shifts caused by the Cs - alkaline atoms spin exchange collisions. The paper presents the interaction potentials of the studied pairs of alkaline atoms and calculates the complex spin exchange cross sections for them. Collisions of alkali-metal atoms in the ground state are considered in the energy interval of 10^{-4}–10^{-2} a.u.

Keywords: alkali atoms, cross sections, spin exchange

Introduction

In recent years, interest has been aroused in the study of the interaction between alkali-metal atoms. This is true for the study of both homonuclear and hetero nuclear dimers of alkaline atoms. The growth of such studies was due to various factors, in particular, interest in the study of "cold" alkaline dimers in traps (Hawamdeh, 2022) of different types and as well as studying the possibility of using alkaline metal as working media in quantum magnetometers and frequency standards (Petrenko, 2021). In particular, taking into account the possibilities of "indirect" optical orientation.

The study of interaction involving spin polarized atoms is possible in experiments on the optical orientation of atoms. The optical orientation of atoms is the transfer of angular momentum from polarized resonant radiation to an ensemble of atoms that are either in the ground state or in an excited state and have an uncompensated electron spin (Happer, 1972).

Collisions of alkali-metal atoms in the ground state with the electron spin S = 1/2 are accompanied by exchange of electron coordinates between the colliding particles, which leads to the polarization transfer between them (i.e., to the well-known phenomenon of spin exchange). In addition, along with the polarization transfer from one partner to another, the magnetic resonance lines of colliding atoms are broadened and shifted in spin exchange collisions. The last two processes depend, in particular, on the complex spin exchange cross section. The spin exchange process in the general case can be described with the help of the complex cross section of the spin exchange. The real part of the cross section determines the so-called "spin exchange cross section," which is responsible for broadening of magnetic resonance lines, while the imaginary part – the shift cross section – governs the frequency shift of magnetic resonance. The study of interactions involving spin polarized atoms began quite a long time ago, but in recent years, interest in these interactions has again become topical. This is due to both the development of experimental technique, in particular, semiconductor lasers, including vertical cavity surface-emitting lasers (VCSEL), which, with their miniature size, allow to achieve a high degree of polarization of alkaline atoms, in particular, cesium. This kind of miniaturization made it possible to consider the possibilities of designing devices, the principle of operation of which is based on the optical orientation of atoms. Such devices include quantum gyroscopes, magnetoencephalographs, and quantum magnetometers.

The method of optical orientation of atoms makes it possible to obtain spin polarized optical particles. The presence of polarized atoms in the working chambers of the above devices leads to the fact that collisions between atoms of the working mixtures make it possible to obtain spin polarized atoms that did not interact with the pumping light (indirect optical orientation of atoms). The paper considers the interaction of spin polarized cesium atoms with alkali atoms Cs, Rb, K, and Li in the ground state.

It should be noted that, in collisions of alkaline atoms at not too low temperatures at which the time of hyperfine interaction $(2\pi/\Delta\omega)$ (for example, $\Delta v = 9192 \times 10^6$ Hz for ^{133}Cs (Radtsig and Smirnov, 1980)) is considerably longer than the collision time, which is of the order of 10^{-12} s, the spin-exchange process can only be considered to be evolution of the electron spins during the collision. In other words, it is assumed that the total electron spin is conserved in the collision process. Coupling of the electron and nuclear spins takes place between the collisions. In this case, the molecule formed from two alkaline atoms in the collision process can be described in the ground state with the help of two potentials corresponding to the total spins of the system $S_1 = 0$ and $S_2 = 1$.

In order to describe the spin-exchange process we have to know complex spin-exchange cross sections. Complex spin-exchange cross sections are calculated on the basis of the data on the singlet ($X^1\Sigma^+$) and triplet ($a^3\Sigma^+$) potentials describing the interaction of alkaline atoms in the ground state.

Spin Exchange Complex Cross Sections

It is known that, when two atomic particles with nonzero electron spins collide, electron exchange process can occur and, if one of the particles was initially polarized, the particles can exchange electron polarization. This process can conditionally be represented as follows:

$$A(\uparrow) + B(\downarrow) \rightarrow (AB) \rightarrow A(\downarrow) + B(\uparrow). \tag{1}$$

The arrows show here possible electron polarization of the particles.

The spin exchange process can be described in terms of a complex cross section of the spin exchange:

$$q^{AB} = \bar{q}^{AB} + i \cdot \bar{\bar{q}}^{AB}. \tag{2}$$

Real part of a cross section determines the orientation transfer in a collision of particles, the relaxation, and the formation of higher polarization moments (alignment, hyperfine polarization) (Dmitriev et al., 1994). Imaginary part of a cross section determines the shifts of a magnetic resonance frequency in the system of both Zeeman and hyperfine levels of atoms. Consequently, knowledge of a spin exchange complex cross section allows to completely describe the processes occurring during spin exchange collisions. The complex cross section of the spin exchange can be conventionally represented in terms of the scattering matrix (Mott and Massey, 1965):

$$q^{AB} = \frac{\pi}{k_{AB}^2} \sum_{l=0}^{\infty} (2l+1) \cdot \left[1 - T_0^{AB}(l) \cdot T_1^{AB}(l)^*\right] \tag{3}$$

here, $k_{AB}^2 = \mu_{AB} \cdot v_{AB}/\hbar$ is the wave vector, μ_{AB} is the reduced mass of colliding particles A and B, v_{AB} is the mean relative velocity of colliding atoms, and the asterisk * denotes complex conjugation. The scattering matrix can be represented in terms of scattering phases ($\delta_S^{AB}(l)$) in a channel with total spin S as follows:

$$T_S^{AB}(l) = \exp(2i\delta_S^{AB}(l)), \tag{4}$$

where l is the orbital quantum number.

From expression (4), it follows that the real and imaginary parts of a complex cross section have the form (Mott and Massey, 1965):

$$\bar{q}^{AB} = \frac{\pi}{k_{AB}^2} \sum_{l=0}^{\infty} (2l+1) \sin^2\left[\delta_1^{AB}(l) - \delta_0^{AB}(l)\right], \tag{5}$$

$$\bar{\bar{q}}^{AB} = \frac{\pi}{k_{AB}^2} \sum_{l=0}^{\infty} (2l+1) \sin 2\left[\delta_1^{AB}(l) - \delta_0^{AB}(l)\right]. \tag{6}$$

According to formulas (5) and (6), the real and imaginary parts of a complex cross section of the spin exchange (2) can be expressed in terms of scattering phases for the singlet and triplet terms of the alkali molecule. The scattering phases were determined in the Jeffrey quasi-classical approximation modified by Langer (Mott and Massey, 1965). The use of the quasi-classical approximation in calculating the scattering phases is quite justified because, in the case of alkali-metal dimers with large reduced mass μ_{AB}, the centrifugal barrier $\left(\frac{(l+1/2)^2}{2\mu_{AB}R^2}\right)$ changes slowly with increasing orbital quantum number l as compared to the kinetic energy. As a result, it is necessary to take into account the contributions of a large number of partial waves to cross sections (5), (6).

Calculation of Complex Spin Exchange Cross Sections

As can be seen from the previous section, the real and the imaginary parts of the complex spin exchange cross section (2) can be expressed by the scattering phases at the singlet and triplet terms of the alkaline dimer in correspondence with formulas (5) and (6). The scattering phases were determined by the Jeffreys approximation modified by Langer as follows (Mott and Massey, 1965):

$$\delta_l = \int_{R_0}^{\infty} F_1(R)dR - \int_{R_0'}^{\infty} F_0(R)dR \quad (7)$$

where

$$F_1^S(R) = \left[2\mu(E - V_S(R)) - \frac{(l+1/2)^2}{2\mu R^2}\right]$$

$$S = s, t, q \quad (8)$$

$$F_0(R) = \left[2\mu E - \frac{(l+1/2)^2}{R^2}\right]$$

here, E is the collision kinetic energy; R_0 and R_0' are roots of the equations $F_1^S(R) = 0$ and $F_0(R) = 0$ (for F_1^S, the largest root should be taken), and $V_S(R)$ is the atomic interaction potential corresponding to the total spin S (for the singlet ($S = 0$) or triplet ($S = 1$) terms) of a system.

For the further use of the obtained values of the cross sections (e.g., in experiments on optical orientation of atoms), it is necessary to pass from the dependences of the cross sections on the collision energy to the dependences on temperature. For this purpose, the cross sections were averaged over the Maxwellian distribution of velocities according to the expression

$$\sigma^{AB}(T) = \frac{\langle q^{AB}(E) v_{AB} \rangle}{\langle v_{AB} \rangle} = \frac{1}{(kT)^2} \int_0^{\infty} q^{AB}(E) E \exp(-E/kT) dE \quad (9)$$

where E is the kinetic energy of colliding atoms, k is the Boltzmann constant, T is the temperature, $q_{AB}(E)$ and $\sigma_{AB}(T)$ are the dependences of cross sections on the kinetic energy and temperature, and v_{AB} is the relative velocity of colliding particles.

Interaction between Spin Polarized Cs and Rb Atoms

Interaction Potentials of Cs–Rb

Collisions of alkaline atoms occur at temperatures at which the hyperfine interaction time is significantly shorter than the collision time, which is about 10^{-12} s; therefore, the spin exchange process can be considered as the evolution of electron spins at the moment of collision. Then, in the interval between collisions, the electronic polarization is redistributed between the electronic and nuclear degrees of freedom of the alkali atom. The molecule formed in a collision process from two alkaline atoms can be described in the ground state in the standard way by two terms corresponding to the total electron spins of the system $S_1 = 0$ and $S_2 = 1$. Consequently, it is necessary to know the singlet ($X^1\Sigma^+$) and triplet ($a^3\Sigma^+$) terms of a dimer, corresponding to the total spins $S_1 = 0$ and $S_2 = 1$, respectively, in order to describe the system of two colliding alkaline atoms in terms of spin-exchange complex cross sections.

In (Docencko et al., 2011, Strauss and Takekoshi, 2010, Pashov et al., 2007) the singlet and triplet terms of CsRb alkali molecules were studied simultaneously by the Fourier spectroscopy method and the interaction potentials of interest to us were constructed using the experimental data. The functional form of the two Born-Oppenheimer potentials ($X^1\Sigma^+$) and ($a^3\Sigma^+$) is split into three regions on the internuclear separation axis R:

1. the short-range repulsive wall ($R < R_{SR}$),
2. the asymptotic long-range region ($R > R_{LR}$),
3. the intermediate deeply bound region in between ($R_{SR} \leq R \leq R_{LR}$).

The potentials obtained in (Docencko et al., 2011, Strauss and Takekoshi, 2010, Pashov et al., 2007) were represented in the following form:

in the short range region ($R < R_i$),

$$U_{SR}(R) = A + \frac{B}{R^q},\tag{10}$$

in the intermediate range region ($R_i < R < R_o$)

$$U_{IR}(R) = \sum_{k=0}^{n} a_k x^k,\tag{11}$$

where $x = \dfrac{R - R_m}{R + bR_m}$. Here the quantities a_i are fitting parameters, in the long range region ($R > R_o$),

$$U_{LR} = -\frac{C_6}{R^6} - \frac{C_8}{R^8} - \frac{C_{10}}{R^{10}} \pm E_{ex},\tag{12}$$

here C_6, C_8, and C_{10} are the van der Waals coefficients. The explicit form of parameters entering into (10)–(13) is presented in (Docencko et al., 2011, Strauss and Takekoshi, 2010, Pashov et al., 2007). Exchange interaction E_{ex} appearing in (12) can be represented in the following form:

$$E_{ex} = A_{ex} R^\gamma \exp(-\beta R),\tag{13}$$

This term appears with minus and plus signs in the expressions for the singlet and triplet terms, respectively.

The singlet and triplet interaction potentials derived in (Docencko et al., 2011) are presented, in the atomic system of units, in Figure 1. The interaction potentials obtained earlier (Kartoshkin, 1995) for the system considered are also shown in the Figure 1. It is evident from the presented plots that the singlet terms obtained in (Docencko et al., 2011, Kartoshkin, 1995) agree rather well, while the triplet potentials differ significantly. In particular, both potential well depths T_e and the positions of minima of the potential energy are different in the case of triplet terms. This is apparently explained by the fact that the asymptotic formulas from (Smirnov, 1973) were used in (Kartoshkin, 1995) to calculate exchange interaction E_{ex}. However, these formulas are valid at long range, but their accuracy decreases considerably with decreasing internuclear separation. Consequently, using the interaction potentials reported in (Docencko et al., 2011) for the singlet and triplet terms in the form (10)–(13), it is possible to calculate the spin exchange complex cross sections for the system in question at both high and low collision energies.

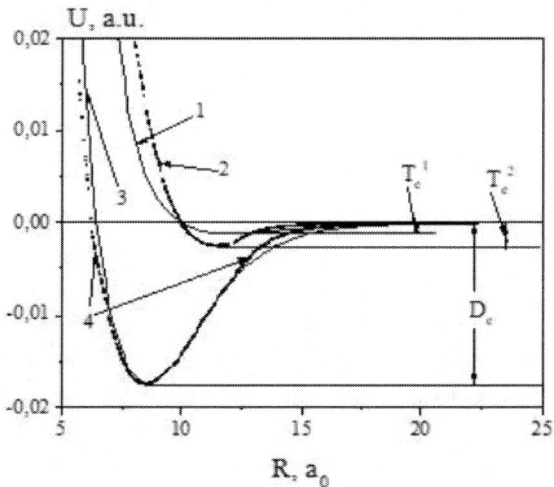

Figure 1. Interaction potentials calculated according to the data of (Kartoshkin, 1995) and (Docencko et al., 2011) : triplet (*1, 2*) and singlet (*3, 4*) from (Docencko et al., 2011), (*1, 3*) and (Kartoshkin, 1995), (*2, 4*). $D_e = -1.747966 \times 10^{-2}$ a.u. (Docencko et al., 2011), $D_e = -1.746429 \times 10^{-2}$ a.u. (Kartoshkin, 1995), $T^1_e = -1.18163539 \times 10^{-3}$ a.u. (Docencko et al., 2011), and $T^2_e = -2.27815 \times 10^{-3}$ a.u. (Kartoshkin, 1995).

Spin Exchange and Frequency Shift Cross Sections

According to formulas (5) and (6), the real and imaginary parts of a complex cross section of the spin exchange (2) can be expressed in terms of scattering phases for the singlet and triplet terms of the CsRb molecule. The scattering phases were determined in the Jeffrey quasi-classical approximation modified by Langer (Mott and Massey, 1965).

Figure 2 shows the dependences of real and imaginary parts of the spin-exchange complex cross section on the collision energy (Kartoshkin, 2016). One can see in the Figure that both cross sections oscillate as the collision energy increases, while the shift cross section becomes positive in the region of an energy value of 0.003 a.u., but remains negative at other energy values.

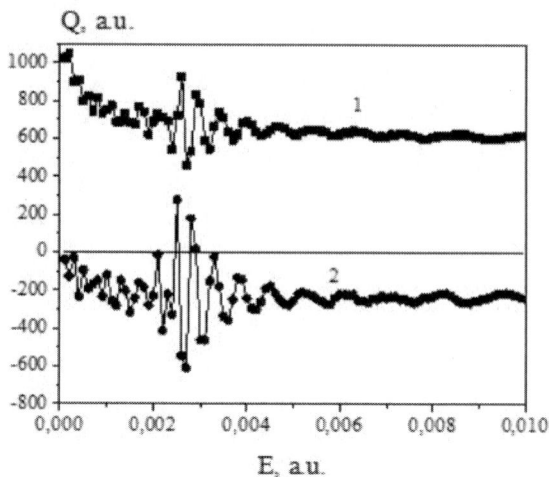

Figure 2. Dependence of (*2*) imaginary and (*1*) real parts of the spin-exchange complex cross section on the collision energy for the CsRb dimer (Kartoshkin, 2016).

Interaction between Spin Polarized Cs Atoms

Interaction Potentials of Cs$_2$

In this study, we used in our calculations the data on the singlet ($X^1\Sigma^+$) state of the Cs$_2$ molecule obtained in (Amiot and Dulieu, 2002). Figure 3 shows the dependences of the energy of interaction in state $X^1\Sigma^+$ on the nuclear spacing. The curve was obtained from the data on the energy of vibrational levels (v = 0 - 135, which corresponds to energies 20.98178 and 3622.03814 cm^{-1} and encompasses the range of nuclear spacings from 3.493792×10^{-8} to 11.037142×10^{-8} cm) tabulated in (Amiot and Dulieu, 2002). Dissociation energy D_e ($X^1\Sigma^+$) = 3649.884 cm^{-1} (Jenc and Brandt, 1989) used in (Xie et al., 2009). In further computations for nuclear spacings smaller than 3.493792×10^{-8} cm, the data on the singlet potential from (Kartoshkin, 1995) were used. Figure 3 also shows the dependences of the triplet interaction potential on the nuclear spacing, which were obtained in (Xie et al., 2009). In the (Xie et al., 2009), the interaction potentials was given in analytic form. The energy of dissociation of the triplet term from (Xie et al., 2009) is T_e = 279.349 cm^{-1}. In addition, the value of D_e used in (Amiot and Dulieu, 2002) determined the dissociation limit for the singlet and triplet terms of dimer Cs$_2$, which is equal to 3650.0321 cm^{-1}. We calculated the spin exchange cross sections of interest using the singlet and triplet terms from (Amiot and Dulieu, 2002) and (Xie et al., 2009), respectively. In both cases, the total energy D_e = 3650.0321 cm^{-1} of dissociated limit was used. It is necessary to use the same value of D_e when the nuclear spacing tends to the range of high values to avoid the error that appears in calculations of cross sections because the value of $\Delta = V_t - V_s$ does not tend to zero.

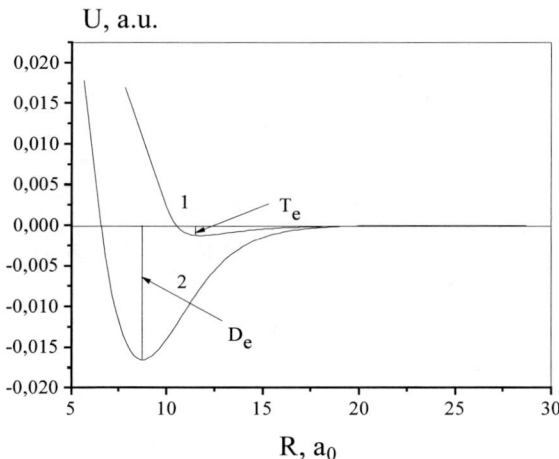

Figure 3. Interaction potentials of dimer Cs$_2$ calculated according to the data of (Amiot and Dulieu, 2002) and (Xie et al., 2009): (*2*) singlet potential obtained in (Amiot and Dulieu, 2002); (*1*) triplet potential obtained in (Xie et al., 2009); D_e = 3650.0321 cm^{-1} and T_e = 279.349 cm^{-1} (Xie et al., 2009).

Spin Exchange and Frequency Shift Cross Sections

The calculation of the cross sections presented in Figure 4 was carried out in accordance with the methodology described above in this chapter, using the interaction potentials shown in Figures 3. Figure 4 shows the dependences of the spin exchange and magnetic resonance frequency shift cross sections on the energy of collision, which were calculated using the formulas of paragraphs 2 and 3 and interaction potentials from (Amiot and Dulieu, 2002, Xie et al., 2009).

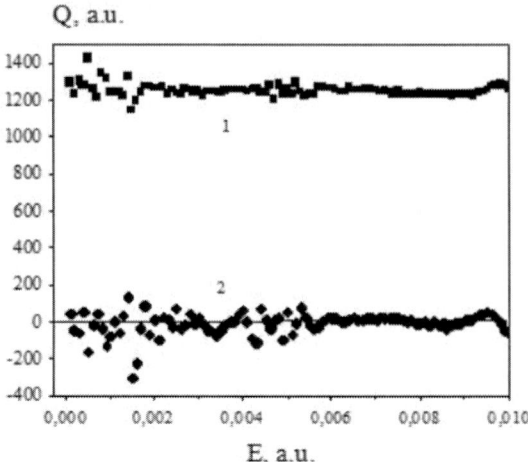

Figure 4. Dependences of the real (1) and imaginary (2) parts of the complex spin exchange cross section on collision energy for the Cs_2 dimer, according to data in (Kartoshkin, 2015).

As seen from Figure 4, the cross sections practically do not change with increasing collision energy. The magnetic resonance frequency shift cross section slightly fluctuates around zero.

Interaction between Spin Polarized Cs and K Atoms

Interaction Potentials of Cs–K

As noted above, when a mixture of two alkali metal atoms is present in the absorption chamber, collisions between similar and different alkali metal atoms occur. Such collisions affect both the redistribution of polarization between colliding atoms (provided that polarized atoms are present in the system, which is the case upon optical orientation of atoms in the volume of an absorption chamber) and the widths and shifts of the magnetic resonance lines of these atoms. In our case such collisions involve the K and Cs atoms. As noted earlier, to describe the influence of spin exchange collisions on the widths and shifts of lines, it suffices to know the complex spin exchange cross sections that are determined by the interaction potentials of colliding atoms. Since the alkali metal atoms in the ground state have the electron spin $S = 1/2$, their collisions give rise to the formation of an alkalimetal dimer, which can be described using the triplet and singlet terms. The singlet and triplet terms correspond, respectively, to the total spin S = 0 and S = 1 of the system. The temperature dependences of the complex spin exchange cross section for the Cs_2 dimer were calculated in (Dmitriev, 2015), and calculations for collisions of two K atoms were carried out in (Kartoshkin, 2011). Consequently, to obtain complete information on the spin exchange collisional processes in a Cs–K mixture, it is necessary to obtain the temperature dependences of complex spin exchange cross sections for these atoms. To calculate the spin exchange cross sections for the K and Cs atoms, we will use the data on the interaction potentials presented in (Ferber et al., 2008, Ferber et al., 2009). Spectroscopic studies of the electronic singlet ($X^1\Sigma^+$) and triplet ($a^3\Sigma^+$) states of the KCs dimer were performed by the Fourier spectroscopy method. The potentials of interest to us were represented analytically in the standard manner in accordance with formulas (14)-(16):

1. the short-range repulsive wall ($R < R_i$),
2. the asymptotic long-range region ($R > R_0$),
3. the intermediate deeply bound region in between ($R_i \leq R \leq R_0$).

in the short range region ($R < R_i$),

$$U_{SR}(R) = A + \frac{B}{R^q}, \tag{14}$$

in the intermediate range region ($R_i < R < R_0$)

$$U_{IR}(R) = \sum_{k=0}^{n} a_k x^k, \tag{15}$$

where $x = \dfrac{R - R_m}{R + bR_m}$. Here the quantities a_i are fitting parameters, in the long range region ($R > R_o$),

$$U_{LR} = -\frac{C_6}{R^6} - \frac{C_8}{R^8} - \frac{C_{10}}{R^{10}} \pm E_{ex}, \tag{16}$$

here C_6, C_8, and C_{10} are the van der Waals coefficients. The explicit form of parameters entering into (14)–(16) is presented in (Ferber et al., 2009). Exchange interaction E_{ex} appearing in (16) can be represented in the following form:

$$E_{ex} = A_{ex} R^{\gamma} \exp(-\beta R), \qquad (17)$$

where, $R_i = 3.185$ Å, $R_0 = 12$ Å, $R_m = 4.28377150$ Å, $q = 6.18977$, $k = 25$, and $b = -0.39$ (for the singlet term) and $R_i = 5.045$ Å, $R_0 = 12.010$ Å, $R_m = 6.04984692$ Å, $q = 3.06025$, $k = 19$, and $b = -0.43$ (for the triplet term). The explicit forms of coefficients A, B, a_k, and C_i are presented in [10]. The term ±Vex that describes the exchange interaction in (3) has the sign "+" in the expression for the triplet term and the sign "–" in the expression for the singlet term.

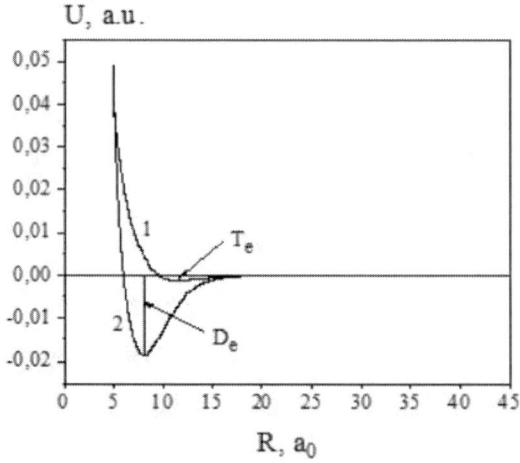

Figure 5. Singlet and triplet interaction potentials of dimer Cs-K calculated according to the data of (Ferber et al., 20090: (2) singlet potential; (1) triplet potential; $D_e = 4069,208$ cm^{-1} and $T_e = 267,141$ cm^{-1} (Ferber et al., 2009).

The dissociation energies of the interaction potentials obtained in (Ferber et al., 2008, Ferber et al., 2009) equal $D_e = 4069.208(40)$ cm^{-1} and $T_e = 267.141(20)$ cm^{-1} for the singlet and triplet terms, respectively

Spin Exchange and Frequency Shift Cross Sections

Figure 6 shows the calculated dependences of real and imaginary parts of the complex spin exchange cross section on the collision energy for system K–Cs taken from (Kartoshkin, 2012). The cross sections for the ^{39}K and ^{133}Cs isotopes were calculated in the range of collision energies from 0.0001 to 0.01 a.u., which corresponds to the temperature interval 30–3000 K. In the energy range studied, the real part of the cross section weakly oscillates around a value of 800 a_0^2 (where a_0 is the Bohr radius), while the imaginary part of the cross section oscillates around a zero value, which is indicative of an insignificant contribution to the magnetic resonance shift for the system in question. To use the cross sections obtained in subsequent calculations or compare them to the experimental data, it is necessary to perform Maxwellian averaging of cross sections over velocities.

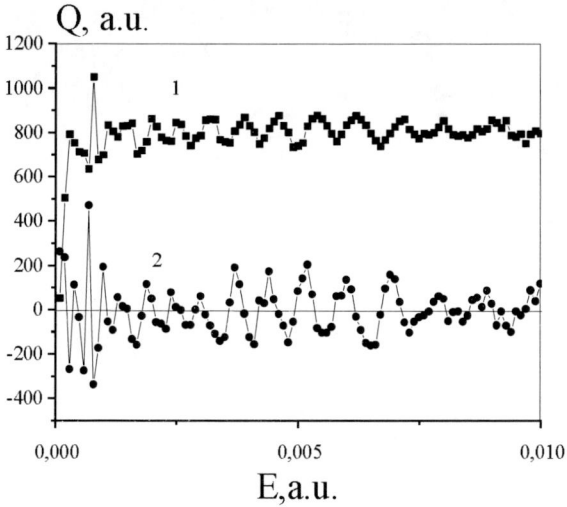

Figure 6. Dependence of the real (1) and imaginary (2) parts of the complex spin exchange cross section on collision energy, according to data in (Kartoshkin, 2012).

Interaction between Spin Polarized Cs and Li Atoms

Interaction Potentials of Cs–Li

Collisions of Cs and Li alkali atoms in the ground state lead to the formation of LiCs molecules. Since the electron spins of the ground-state colliding atoms are 1/2, the formed molecule can exist in two states, i.e., with total spins S equal to 0 (singlet state) and 1 (triplet state). The singlet and triplet terms describing the interaction between Cs and Li atoms in the ground state were presented in (Staanum et al., 2007). Using high-resolution Fourier spectroscopy, the authors of (Staanum et al., 2007) obtained and tabulated data for the singlet and triplet interaction potentials of the $^7\text{Li}^{133}\text{Cs}$ dimer.

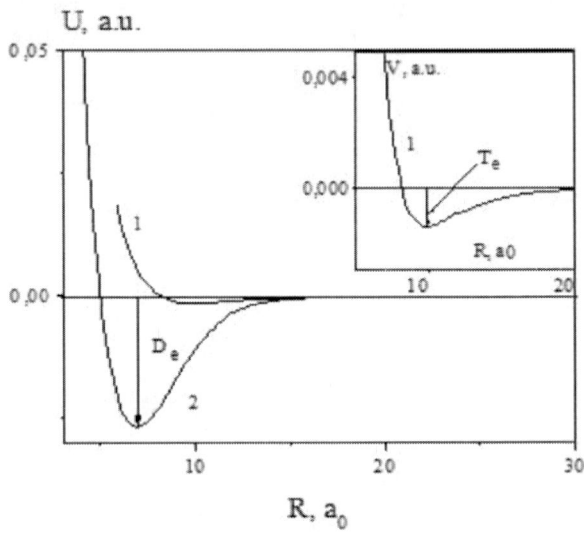

Figure 7. Singlet (2) and triplet (1) interaction potentials of $^7\text{Li}-^{133}\text{Cs}$ system calculated according to the data of (Staanum et al., 2007). The inset shows the triplet potential on an enlarged scale.

Figure 7 shows the interaction potentials for the singlet (2) and triplet (1) terms of the $^{133}\text{Cs}^7\text{Li}$ dimer in the atomic system of units. Work (Staanum et al., 2007) gives the following dissociation energies D_e and equilibrium distances R_e for these potentials: D_e = 5875.455 cm^{-1} (0.0267627 a.u.) and R_e = 3.66681 Å (6.932 a_0) for the singlet potential and T_e = 309 cm^{-1} (0.0014075 a.u.) and R_e = 5.2472 Å (9.919, a_0) for the triplet potential. Using these data, we can calculate the complex spin exchange cross sections and use them to determine the magnetic resonance frequency shifts in collisions of alkali atoms.

Spin Exchange and Frequency Shift Cross Sections

Knowing the singlet and triplet interaction potentials of the LiCs molecule, we can calculate the phases of scattering from the singlet and triplet terms and determine the cross sections of interest.

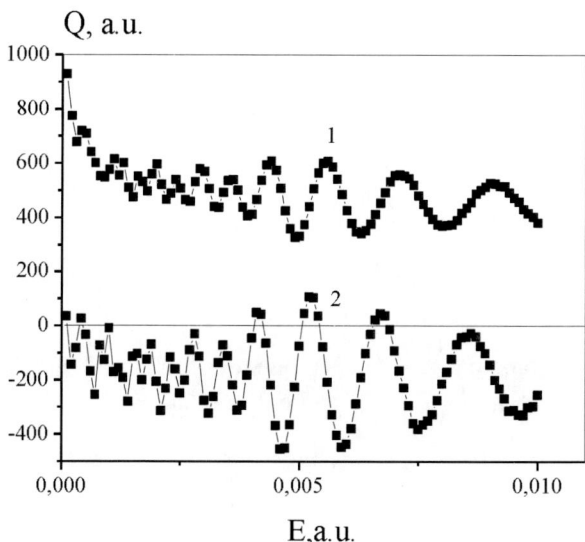

Figure 8. Dependence of the (1) real and (2) imaginary parts of the complex spin exchange cross section on the collision energy in the system of two ^{7}Li–^{133}Cs atoms (in the atomic system of units) (Kartoshkin, 2020).

The scattering phases were calculated in the quasi-classical approximation in the collision energy range of 10^{-4}–10^{-2} a.u. using the Jeffreys approximation modified by Lambert.

Figure 8 shows the calculated dependences of the real (1) and imaginary (2) parts of the complex cross section on the collision energy. One can see that the frequency shift cross section of LiCs lies mainly in the region of negative values.

As can be seen from the Figure, both cross sections oscillate over the entire range of energies under study.

Thus, the temperature dependences of the real and imaginary parts of the complex spin-exchange cross section obtained in the present work allow to calculate the contribution of the spin exchange process to the width and shift of the magnetic resonance line upon collisions between polarized lithium and cesium atoms. However, since collisions in the absorption chamber occur not only between Li and Cs atoms, but also between identical atoms (Li–Li, Cs–

Cs), to perform calculations, it is also necessary to know the complex spin-exchange cross sections describing collisions between identical atom.

Spin Exchange Cross Sections

In order to make a comparison with experimental data or data of other authors, it is necessary to pass from the energy dependences of the cross sections presented in Figures. 2, 4, 6, 8 to the dependence of these cross sections on the temperature in the working chambers. Table 1 presents the Maxwell-averaged real and imaginary parts of the complex cross sections for the spin exchange of the systems under study. Interesting for us cross sections present in the temperature range of 300–400 K.

Table 1. Spin exchange (QTR) and frequency shift (QSH) cross sections for systems Cs-Cs, Cs-Rb, Cs-K, and Cs-Li

Systems	Cross section 10^{-16}cm^2	T,K		
		300	350	400
Cs-Cs	QTR,	352	351	350
	QSH,	-6,3	-5,7	-5,1
Cs-Rb	QTR	202	198	196
	QSH,	-60	-62	-63
Cs-K	QTR,	324	327	328
	QSH,	-7,4	-7,2	-6,9
Cs-Li	QTR,	154	151	149
	QSH	-41	-42	-44

Conclusion

As can be seen from the data in Table 1, the real part of the spin exchange cross section (QTR) is quite large in magnitude and lies in the range from 1.5 to 3.5 in units of 10^{-14}cm^2. At the same time, the imaginary part of the cross section, which is responsible for the shifts of the magnetic resonance frequency, is, firstly, negative in magnitude for all the systems under study, and secondly, it varies from 10^{-16}cm^2 to 10^{-15}cm^2 for different systems.

The negative sign of the shift cross section indicates that the interaction of two atoms gives rise to a quasi-molecule, which then decays into the initial atoms, while the frequency of the magnetic resonance of the atoms shifts towards lower values with respect to the exact resonance.

As noted above, the cross sections for spin exchange and frequency shift presented in this chapter make it possible to describe the process of spin exchange in the collision of two alkali atoms in the ground state. The spin exchange process leads both to the transfer of polarization between the colliding particles and to the broadening of the magnetic resonance line of the colliding atoms. In addition, along with the polarization transfer from one partner to another, the magnetic resonance lines of colliding atoms are broadened and shifted in spin-exchange collisions. These processes depend, in particular, on the complex spin-exchange cross sections. The real part of the cross section determines the so-called "spin exchange cross section," which is responsible for broadening of magnetic resonance lines, while the imaginary part - the shift cross section-governs the frequency shift of magnetic resonance. The spin-exchange broadening of a magnetic resonance line affects the precision of such quantum electronics devices as quantum frequency standards and magnetometers.

Disclaimer

None.

References

Amiot C and Dulieu O. The Cs_2 Ground Electronic State by Fourier Transform Spectroscopy: Dispersion Coefficients. *J. Chem. Phys.* (2002) 117(11): 5155-5164.

Dmitriev S P, Dovator N A, Kartoshkin V A, and Okunevich A I. Observation of the Hyperfine Polarization and Alignment in the Spin-exchange Collisions of the Polarized Cs and Rb Atoms. *Optica i Spectroscopia* (1994) 77(5): 712-713.

Dmitriev S P, Dovator N A, Kartoshkin V A. Spin Exchange upon Collision of Two Cesium Atoms in the Ground State. *Tech. Phys.* (2015) 60(6): 826-829.

Docencko O, Tamanis M, Ferber R, Knöckel H, Tiemann E. Singlet and Triplet Potentials of the Ground-state Atom Pair Rb + Cs Studied by Fourier-transform Spectroscopy. *Phys. Rev. A* (2011) 83(5): 052519.

Ferber R, Klincare I, Nikolayeva O, Tamanis M, Knöckel H, Tiemann E, Pashov A. The Ground Electronic State of KCs Studied by Fourier Transform Spectroscopy. *J. Chem. Phys.* (2008)128(24): 244316.

Ferber R, Klincare I, Nikolayeva O, Tamanis M, Knöckel H, Tiemann E, Pashov A. $X^1\Sigma^+$ and a $^3\Sigma^+$ States of the Atom Pair K+Cs Studied by Fourier-transform Spectroscopy. *Phys. Rev. A* (2009) 80(6) 062501.

Happer W. Optical Pumping. *Rev. Mod. Phys.* (1972) 44(2):169-249.

Hawamdeh M M, Sandouqa A S, Joudeh B R, Al-Obeidat O T, Ghassib H B. Scattering Properties of ^{7}Li Vapor. *Eur. Phys. J. Plus.* (2022) 137 (9):1025.

Jenc F, and Brandt B A. Application of the Reduced-potential Curve Method for the Detection of Errors or Inaccuracies in the Analysis of Spectra and for the Construction of Internuclear Potentials of Diatomic Molecules: Alkali diatomic molecules. *Phys. Rev. A* (1989) 39(9):4561-4582.

Kartoshkin V A. Complex Cross Sections for the Spin Exchange between Cs and Rb Alkali Atoms. *Optica i Spectroscopia* (1995) 79(8): 26-31. (in Russian).

Kartoshkin V A. Spin Exchange during Collisions of Potassium Atoms: Complex Cross Sections. *Opt. Spectrosc.* (2011) 111(12):881-884.

Kartoshkin V A. Spin Exchange Processes in a Single-chamber Cs-K Tandem Magnetometer. *Opt. Spectrosc.* (2012) 113(3):235–239.

Kartoshkin V A. Collisions of Alkali-metal Atoms Cs and Rb in the Ground State. Spin Exchange Cross Sections. *Opt. Spectrosc.* (2016) 121(3): 327–330.

Kartoshkin V A. Collisions with Participation of Polarized Cesium and Lithium Alkali Atoms. *Opt. Spectrosc.* (2020) 128(4): 470–472.

Mott N F, Massey H S W. 1965. *The Theory of Atomic Collisions*. Oxford: Clarendon Press.

Pashov A, Docenko O, Tamanis M, Ferber R, Knöckel H, Tiemann E. Coupling of the $X^1\Sigma^+$ and $a^3\Sigma^+$ States of KRb. *Phys. Rev. A* (2007) 76(2): 022511.

Radtsig, A A, Smirnov, B M. 1980. *Handbook of Atomic and Molecular Physics*. Moscow: Atomizdat. (in Russian).

Petrenko M V, Pazgalev A S, Vershovskii A K. Ultimate Parameters of the All-Optical Single-Beam Nonzero Magnetic Field Sensor for Biological Applications. *IEEE Magnetics Letters.* (2021) 12():8104605.

Smirnov, B M 1973. *Asymptotic Methods in the Theory of Atomic Collisions*. Moscow: Atomizdat. (in Russian).

Staanum P, Pashov A, Knöckel H, Tiemann E. $X^1\Sigma^+$ and $a^3\Sigma^+$ States of LiCs Studied by Fourier-transform Spectroscopy. *Phys. Rev. A* (2007) 75(4): 042513.

Strauss C, Takekoshi T, Lang F, Winkler K. Hyperfine, Rotational, and Vibrational Structure of the $a^3\Sigma^+_u$ State of ^{87}Rb$_2$. *Phys. Rev. A* (2010) 82(5): 052514.

Xie F, Sovkov V B. Lyyra A M, Li D, Ingram S, Bai J, Ivanov V S, Magnier S, Li L. Experimental Investigation of the Cs$_2$ a $^3\Sigma^+_u$ Triplet Ground State: Multiparameter Morse Long Range Potential Analysis and Molecular Constants. *J. Chem. Phys.* (2009) 130(5): 051102.

Biographical Sketch

Victor Kartoshkin

Affiliation: Ioffe Institute, Division of Plasma Physics, Atomic Physics and Astrophysics, Lab. of Atomic Radiospectroscopy, Russian Federation.
Education: In 1975 I graduated from the Radiophysics Faculty of the M. I. Kalinin Leningrad Polytechnic Institute (Now it is Peter the Great St. Petersburg Polytechnic University).

Business Address: 26 Politekhnicheskaya, St Petersburg 194021, Russian Federation

Research and Professional Experience
After graduating from the University in 1975 and to the present day, I am working at the Ioffe Institute. In 1980 I defended my Ph.D. thesis on the topic «Investigation of the process of metastability exchange in a mixture of helium isotopes He3-He4». In 1990 I defended my doctoral dissertation «Atom-Atomic and Atom-Molecular Interactions Involving Excited Polarized Atoms of Light Inert Gases». My scientific interests are focused on the study of collisional processes between spin-polarized atomic and molecular particles. Study of elastic and inelastic processes at the collision between such particles, transfer and conservation of spin polarization in such processes. The use of the results obtained in the creation of quantum electronics devices based on the principles of optical orientation of atoms. I have over 100 publicaations in peer-reviewed journals. The results of my research have been presented in more than 100 reports at various international conferences. I was the leader of a number of scientific projects in the field of applied and fundamental researches.

Professional Appointments: Chief Research Officer

Honors: International Society for Magnetic Resonance, member

Publications from the Last 3 Years
[1]   Kartoshkin, V A. Angular Momentum Transfer at the Chemo-ionization and Spin Exchange Processes. Interaction between Spin Polarized

Metastable Helium Atoms//2022, Mediterran. *J. Chem.*, v.12 (2), 164-174.

[2] Kartoshkin, V A. Сдвиги частоты магнитного резонанса щелочных атомов в смеси K-Li (Magnetic resonance frequency shifts of alkali atoms in the K-Li mixture.) 2022, *Оптика спектроск.*, т.130 (11),1634-1637 DOI: https://doi.org/10. 21883/OS.2022.11.53767. 3522-22 (in Russian).

[3] Kartoshkin V A. Collisions of Lithium Atoms in Ground State. Complex Spin-Exchange Cross Sections //2021, *Opt. Spectrosc.*, v.129 (6), 641-644.

[4] Kartoshkin, V A. Magnetic Resonance Frequency Shift of Na Atoms during Collisions in a Mixture of Potassium and Sodium Atoms // 2021, *Tech. Phys.*, v.66 (11), 1221-1227.

[5] Kartoshkin, V A. Magnetic-Resonance Frequency Shifts in a Tandem Cs-K Magnetometer Induced by Spin Exchange // 2020, *Opt. Spectrosc.*, v.128 (9),1355-1358.

[6] Kartoshkin, V A. Collisions with Participation of Polarized Cesium and Lithium Alkali Atoms //2020, *Opt. Spectrosc.*, v.128 (4), 470-472.

[7] Kartoshkin, V A. Frequency shifts of the magnetic resonance of Rb and K atoms in the K-Rb tandem magnetometer // International Conference PhysicA.SPb/2020 *J. Phys.*: Conf. Ser., v.1697 (1), 01214.

[8] *International conference PhysicA.SPb/2019.* 19-23 October 2020, Saint-Petersburg, Russia. Kartoshkin V. A. Frequency shifts of the magnetic resonance of Rb and K atoms in the Rb-K tandem magnetometer. P. 241.

[9] *International conference PhysicA.SPb/2019.* 18-22 October 2021, Saint-Petersburg, Russia. V. A. Kartoshkin. Frequency shift of the magnetic resonance at the spin exchange collisions between K and Li atoms. P. 3-30.

[10] *The Virtual 32nd International Conference on Photonic, Electronic and Atomic Collisions (ViCPEAC 2021)* from July 20-23, 2021. Canada. V. Kartoshkin. Transfer of angular momentum at the interaction between spin polarized metastable helium atoms. P. 107.

[11] *On line 27th International Symposium on Ion-Atom Collisions (27 ISIAC)* July 14-16, 2021. Romania. V. Kartoshkin, S. Dmitriev, N. Dovator. Ionization of alkali atoms during the collisions with polarized metastable helium. Redistribution of the spin polarization. P. 30.

[12] *International conference PhysicA*.SPb/2022. 17-21 October 2022, Saint-Petersburg, Russia V. Kartoshkin. Collisions of spin polarized Cs atoms with alkali atoms. Spin exchange cross sections and magnetic resonance frequency shifts. P. 217.

Chapter 4

Superconducting State Parameters of Alkali Metals

Rajesh C. Malan[1,*]
and Aditya M. Vora[2,†]

[1]Applied Science and Humanities Department,
Government Engineering College Valsad, Gujarat, India
[2]Department of Physics,
University Schools of Science, Gujarat, India

Abstract

> The current chapter is presented with an aim to investigate screening dependency of the superconducting behaviour of alkali metals. The screening dependent superconducting state parameters for five alkali metals (Li, Na, K, Rb, and Cs) are computed in the present work. The pseudopotential given by Fiolhais and his co-worker along with its universal parameters is used for the entire calculation of superconducting state parameters. Critical/transition temperature (T_c), effective interaction strength (N_0V), electron-phonon coupling strength (λ) and Coulomb pseudopotential (μ) are the superconducting state parameters included in the present study. Very weak superconducting behaviour is observed for alkali metals. The screening and exchange and correlation effect dependency of the present results is also discussed in the chapter.

---

[*]Corresponding Author's Email: rcmgecv@gmail.com
[†]Corresponding Author's Email: voraam@gmail.com

In: Alkali Metals
Editor: Wilbur M. Hulett
ISBN: 979-8-88697-706-6
© 2023 Nova Science Publishers, Inc.

PACS 74.70.-b, 74.25.-q, 74.20.Pq.

Keywords: superconducting state parameters, pseudopotential theory, alkali metals

1. Introduction

In conjunction with our comprehensive study of alkali metals and alloys [1–7] using pseudopotential theory, few screening dependent superconducting state parameters are calculated for the solid phase of alkali metals (Li, Na, K, Rb and Cs). Pseudopotential theory is one of the successful tools for the study of superconductivity in metals as well as alloys.

Being a very sensitive properties, the experimental study dealing with superconducting properties seeks extensive accuracy with detailed relevant database. Thus, there is a strong demand for accurately predicted information prior to start the experimental research. Hence, the theoretical results proved a hidden backbone of the superconductivity research.

The superconducting state of any material can be characterized by the superconducting state parameters such as temperature of the phase transition from normal state to superconducting state (critical/transition temperature (T_c), Coulomb pseudopotential (μ), electron-phonon mass enhancement (Γ), isotopic effect exponent (α), effective interaction strength (N_0V) and electron-phonon coupling strength (λ). Because of the dependency of these parameters on electron-ion interaction, pseudopotential has a sound ability to estimate them. Many researchers have reported the calculation of these parameters for metals using pseudopotential theory [8–19]. Khanna et al. [8] calculated the above parameters along with other properties of the five metals (Li, Na, K, Al and Pb). The solution of the Eliashberg equation [9] has been used to obtain the superconducting state parameters [10]. Jain et al. [11] have reported these parameters for metals. Allen and Cohen [12] have included more than fifteen different metals to obtain Γ and T_c using the pseudopotential and phonon spectra. Apart from these initiative works, Gajjar et al. [13] have provided the calculation for all of these parameters for monovalent, divalent and polyvalent metals. It can be easily seen from the above work that the pseudopotential can be a good theoretical method to obtain the said parameters for any metallic superconductor. For the alloys and compound materials some mathematical formulations required to be introduced in the existing theories of metals. The first successful attempt was made by Allen and Dynes [14] to present the parameters of Pb, In and Tl based binary alloys by extending BCS theory up to alloys. The superconducting parameters of the In based binary alloys has been calculated by Khan et al. [15] and Singh et al. [8] separately. With the development in computing hardware and a deep understanding of the phenomenon, researchers are now able to provide the predictions regarding superconductivity. Many researchers have produced the superconducting state parameters of even binary and tertiary alloys using the pseudopotential theory [14–19]. Different model potentials generate superconducting state parameters with different accuracy. In the case of alkali metals and alloys, the previous calculations by different researchers show very poor or almost no superconducting behaviour of alkali metals and alloys [14,20,21]. To meet the two goals of the present research [(i) study of screening dependency of superconducting state parameters of alkali metals and (ii) test the suitability of model potential of Fiolhais et al. [22, 23] for calculation of superconducting state parameters], five alkali metals are included in the current study. The potential given by Fiolhais et al. [22, 23] in universal form has been used. To obtain only natural results by the electron-ion interaction, no parameter fitting has been made. The form of the pseudopotential $V(q)$ is as given below [22, 23],

$$V(q) = 4\pi Z(R_U)^2 \times \qquad (1)$$
$$\times \left[-\frac{1}{(q(R_U))^2} - \frac{1}{(q(R_U))^2 + \alpha_U^2} + \frac{2\alpha_U \beta}{[(q(R_U))^2 + \alpha_U^2]^2} + \frac{2A}{[(q(R_U))^2 + 1]^2} \right]$$

Where the values of the α_U and R_U are provided by Fiolhais et al. [22, 23] and in the present calculation, it is directly adapted to test its behaviour for presenting superconductivity. The exchange and correlation as well as the screening effect also tested over the Fiolhais et al. [22, 23] bare-ion potential using the static Hartree function (H) [24] and the other local field correction functions viz: Hubbard and Shame (HS) [25, 26], Vashishtha-Singwi (VS) [27], Taylor (T) [28], Sarkar et al. (S) [29], Ichimaru-Utsumi (IU) [30], Farid et al. (F) [31] and Nagy (N) [32]. On the bases of our previous successful results for metals as well as alloys for other properties [1–7], the above pseudopotential for the current work has been chosen again. Thus, the empirical formulas for calculation of superconducting state parameters are used along with the pseudopotential theory.

2. Theory

According to McMillan [10], the coupling strength between electron and phonon can be calculated as [13, 26],

$$\lambda = \frac{m^* \Omega}{4\pi^2 k_f M <\omega^2>} \int_0^{2f} q^3 |V(q)^2| dq \qquad (2)$$

Where m^*, Ω, k_f, M and $V(q)$ are effective mass, atomic volume, Fermi wave vector, ionic mass and bare-ion potential respectively. ω^2 is the averaged square of phonon frequency.

To describe the effect of Coulomb potential on the superconductivity, the factor used is known as the Coulomb pseudopotential (μ^*). In the present calculation, the Coulomb pseudopotential (μ^*) is calculated from the following expression [13, 20],

$$\mu^* = \frac{\frac{m_b}{\pi k_f} \int_0^1 \frac{dX}{\epsilon(x)}}{1 + \frac{m_b}{\pi k_f} \ln\left(\frac{E_f}{10\theta_D}\right) \int_0^1 \frac{dX}{\epsilon(x)}} \qquad (3)$$

It can be seen from the above equation that the μ^* involves the Fermi energy (E_F), Debye temperature (θ_D), band mass of the electron (m_b). $\epsilon(X)$ stands for the modified Hartree dielectric function (H). The McMillan's formula correlates the T_c and α with the λ and μ^*. The T_c of an alloy can be obtained as [13, 20],

$$T_c = \frac{\theta_D}{1.45} exp\left[\frac{-1.04(1+\lambda)}{\lambda - \mu^*(1 + 0.62\lambda)}\right] \qquad (4)$$

The effective interaction strength $N_0 V$ can be expressed in terms of λ and μ^* as given below [13, 20]

$$N_0 V = \frac{\lambda - \mu^*}{1 + \frac{10}{11}\lambda} \qquad (5)$$

3. Results and Discussion

The input values of θ_D, z, m_b and m^* are shown in Table 1. The potential parameters α_U and R_U used for the calculations are also shown in Table 1.

The following sections will describe the superconducting state parameters of alkali metals. For the validation of present results, it is compared with the other results [11, 13, 20, 34–36, 38].

Table 1. Input parameters and constants

Metal	θ_D K [20]	z [20]	m_b (au) [20]	m_* (au) [20]	Potential parameters	
					α_U [22,23]	R_U [22,23]
Li	352	1	1.19	2.2	3.549	0.361
Na	157	1	1	1.26	3.075	0.528
K	59.4	1	0.94	1.24	2.807	0.745
Rb	54	1	1	1.25	2.748	0.824
Cs	40	1	0.98	1.43	2.692	0.92

3.1. Electron-Phonon Coupling Strength (λ)

Table 2 shows the obtained results of λ for five alkali metals.

Table 2. λ (au) for alkali metals

Metal	H	HS	VS	T	IU	F	S	N	Others [11,13,34–36,38]
Li	0.433	0.543	0.630	0.670	0.719	0.722	0.586	0.636	0.10, 0.11, 0.21, 1.01, 0.56, 0.45
Na	0.268	0.343	0.411	0.438	0.475	0.477	0.377	0.410	0.260, 0.418, 0.452, 0.12
K	0.460	0.605	0.754	0.811	0.888	0.893	0.679	0.738	0.347, 0.584, 0.635, 0.659, 0.12, 0.14, 0.22
Rb	0.209	0.277	0.348	0.374	0.410	0.412	0.312	0.339	0.390, 0.676, 0.7380, 0.12, 0.14
Cs	0.178	0.238	0.307	0.333	0.367	0.369	0.272	0.295	0.317, 0.592, 0.619, 0.666, 0.484, 0.12, 0.19, 0.28

The origin of the superconductivity lies in the electron-electron interaction via phonon. The cooper pairs are the results of the exceeding interaction strength than the Coulomb repulsion. The electron-phonon coupling (λ) plays an important role in setting the variety of properties including thermodynamical and transport properties, particularly of metals. Successive development after the BCS theory shows the role of electron-phonon coupling by the theory of Eliashberg and the involvement of strong electron-phonon coupling by Migdal-Eliashberg theory [37] in deciding the superconducting state of any material. Thus, considering the importance of the phonon coupling, in the present article, the electron-phonon coupling strength (λ) is also included. As can be seen from Table 2, the results for the Hartree function (H) is found minimum among all correction functions and the results for the F-function are found maximum for all metals. Local field correction functions have a significant effect on λ. The present results are compared with several other results [11, 13, 34–36, 38] for alkali metals. Presently calculated value of λ doesn't exceed 1 for any of the metals, which shows the weaker coupling strength and hence poor superconducting behaviour of alkali metals as compared to the normal-superconducting materials.

3.2. Coulomb Pseudopotential (μ^*)

The Coulomb pseudopotential (μ^*) explains the effect of the Coulomb repulsion on superconductivity. For a normal system, larger the screening effect and retardation, lesser will be the values of μ^*. Depending upon, whether the system is closed or open for the Mott transition, the dominance of the screening effect or Coulomb pseudopotential (μ^*) can be decided, which alternately decides the dependency of superconducting transition temperature (T_c) on pressure.

Table 3 shows the calculated values of μ^* for metals under study. As in the case of calculation of λ, the H-function provides the minimum value of μ^* and maximum for the F-function. The comparison of the current calculation with others results [11–14, 34–36, 38] shows a good agreement. The effect of the local field correction function is not observed considerably large on the value of μ^* for any of the five metals.

Superconducting State Parameters of Alkali Metals 89

Table 3. Coulomb pseudopotential (μ^*)

Metal	H	HS	VS	T	IU	F	S	N	Others [11–14, 34–36, 38]
Li	0.209	0.222	0.231	0.237	0.241	0.241	0.227	0.231	0.183, 0.203, 0.206, 0.207, 0.197, 0.16, 0.17, 0.17, 0.18
Na	0.183	0.194	0.204	0.209	0.213	0.213	0.200	0.203	0.162, 0.182, 0.185, 0.187, 0.177, 0.16
K	0.168	0.178	0.188	0.193	0.196	0.197	0.184	0.186	0.160, 0.183, 0.186, 0.188, 0.177, 0.15, 0.16, 0.18
Rb	0.175	0.186	0.196	0.201	0.205	0.205	0.192	0.194	0.1569, 0.1779, 0.1805, 0.1826, 0.1719, 0.18, 0.15
Cs	0.172	0.183	0.194	0.200	0.203	0.204	0.190	0.192	0.1540, 0.1753, 0.1778, 0.1800, 0.1687

3.3. Transition Temperature (T_c)

The transition temperature (T_c) is one of the key features of any superconducting material as it decides the temperature of transition from the superconducting state to the normal state. Hence, it also gives a direct idea about the required laboratory environment and therefore the cost to obtain the superconducting state also. Presently calculated transition temperature of alkali metals are given in Table 4. It can be observed that for no metal T_c is found more than 2K. This indicates that it is very difficult to convert alkali metals in a superconductor. Present results provide a very low value of T_c for all alkali metals under study. Presently obtained results are also compared with the other theoretical results [11, 13, 38].

Table 4. Transition Temperature (T_c)

Metal	H	HS	VS	T	IU	F	S	N	Others [11, 13, 38]
Li	0.0327	0.359	0.997	1.36	1.96	1.99	0.623	0.623	0.02, 0.02, 1.1, 1.5
Na	3.96×10^{-09}	2.49×10^{-04}	0.0083	0.0186	0.0497	0.0512	0.00191	0.00191	1.47×10^{-6}, 0.043, 0.096, 0.128, 0.014
K	0.082	0.393	0.898	1.10	1.42	1.43	0.625	0.625	0.006, 0.450, 0.666, 0.770, 0.268
Rb	3.13×10^{-46}	5.69×10^{-9}	9.69×10^{-5}	4.36×10^{-4}	2.61×10^{-03}	2.74×10^{-03}	2.49×10^{-06}	2.49×10^{-06}	0.023, 0.607, 0.828, 0.937, 0.396
Cs	7.02×10^{-19}	1.49×10^{-19}	3.94×10^{-7}	7.96×10^{-6}	1.59×10^{-04}	1.76×10^{-04}	8.82×10^{-11}	8.82×10^{-11}	0.0009, 0.252, 0.303, 0.410, 0.081

3.4. Effective Interaction Strength (N_0V)

The effective interaction strength (N_0V) for the metals under study is given in Table 5. For Li, our results estimate N_0V values less than that of the Gajjar et al. [13] (except for Hartree function (H)), whereas for the other alkali metals, the present results are almost near to the results of the others [11, 13, 34, 38].

Table 5. Effective interaction strength (N_0V)

Metal	H	HS	VS	T	IU	F	S	N	Others [11,13,34, 38]
Li	0.160	0.215	0.254	0.269	0.289	0.290	0.234	0.257	0.458, 0.595, 0.616, 0.626, 0.560, 0.13
Na	0.068	0.113	0.151	0.164	0.183	0.184	0.132	0.151	0.080, 0.170, 0.190, 0.197, 0.149, -0.17
K	0.206	0.275	0.336	0.356	0.383	0.384	0.306	0.330	0.143, 0.262, 0.285, 0.294, 0.236, -0.45
Rb	0.029	0.072	0.115	0.129	0.150	0.150	0.093	0.111	0.172, 0.309, 0.336, 0.345, 0.278, -0.005
Cs	0.005	0.045	0.088	0.102	0.123	0.124	0.066	0.081	0.127, 0.271, 0.283, 0.303, 0.219

Concluding Remarks

The superconducting mechanism in five alkali metals has been studied in the present work. The formalism used in the present work is based on electron-phonon coupling strength provided by McMillan [10]. Four superconducting state parameters namely electron-phonon coupling strength (λ), Coulomb pseudopotential ($\mu*$), transition temperature (T_c) and effective interaction strength (N_0V) have been computed in the present article. The transition temperature T_c has found very less uncertain in behaviour as compared to the other results [11, 13, 38]. The potential suggested by the Fiolhais et al. [22, 23] in its original form with universal parameters is found suitable for present calculations. In fact, the obtained results are acceptable for many superconducting parameters even it is obtained without any parameter fitting. Thus, the article provides an opportunity to the researcher for getting better results for other metals as well as alloys for the study of superconductivity with the same model pseudopotential. Very high impact of local field correction functions of VS and T is obtained for all parameters under study. Static Hartree function (H) provides the results nearest to other theoretical predictions for all parameters under the current study.

References

[1] R.C. Malan, A.M. Vora (2019) Study of collective motion in liquid alkali metals. *Bull. Mater. Sci.* **42** 4 1-6 DOI: https://doi.org/10.1007/s12034-019-1836-y.

[2] R.C. Malan, A.M. Vora (2020) Investigation of dynamical properties of liquid alkali metals. *Phys. Chem. Liq.* **58** 2 222-229 DOI: https://doi.org/10.1080/00319104.2018.1564304.

[3] R.C. Malan, A.M. Vora (2019) Thermodynamical Properties of $K_{1-X}Rb_X$ Alloys in Liquid State. *Mater. Today: Proc.* **12** 3 671-675 DOI: https://doi.org/10.1016/j.matpr.2019.03.112.

[4] R.C. Malan, A.M. Vora (2018) Electrical resistivity of liquid Na-alkali alloys. *AIP Conf. Proc.* **1953** 1 140014 DOI: https://doi.org/10.1063/1.5033189.

[5] R. C. Malan, A. M. Vora (2019) Electrical Transport Properties of Liquid $Li_{1?x}Na_x$ Alloys. *J. Nano? Elec. Phys.* **11** 1 01004-1 DOI: doi.org/10.21272/jnep.11(1).01004.

[6] R.C. Malan, A.M. Vora (2018) Thermodynamical investigation of liquid $Li_{1-x}K_x$ alloy. *AIP Conf. Proc.* **2009** 020052-020056 DOI: https://doi.org/10.1063/1.5052121.

[7] R.C. Malan, A.M. Vora (2018) Thermodynamical investigation of liquid alkali metals with Gibbs-Bogoliubov method. *J. Nano- Elec. Phys.* **10** 1 03020 DOI: https://doi.org/10.21272/jnep.10(1).01002.

[8] K.N. Khanna, P.K. Sharma (1979) Binding energy, compressibility, interionic potential, and transition temperature of simple metals obtained by the pseudopotential method. *Phy. Status Sol (b)* **91** 1 251-256 DOI: https://doi.org/10.1002/pssb.2220910126.

[9] G.M. Eliashberg (1960) Interactions between electrons and lattice vibrations in a superconductor. *Soviet. Phys. - J. Exper. Theor. Phys.* **11** 3 696-702.

[10] W.L. McMillan (1968) Transition temperature of strong-coupled superconductors. *Phys. Rev.* **167** 2 331 DOI: https://doi.org/10.1103/PhysRev.167.331.

[11] S.C. Jain, C.M. Kachhava (1980) Electron-Phonon Interaction and Superconductivity in Metals. *Phys. Stat. Sol. (b)* **101** 2 619-626 DOI: https://doi.org/10.1002/pssb.2221010222.

[12] P.B. Allen, M.L. Cohen (1969) Pseudopotential calculation of the mass enhancement and superconducting transition temperature of simple metals. *Phys. Rev.* **187** 2 525 DOI: https://doi.org/10.1103/PhysRev.187.525.

[13] P.N. Gajjar, A.M. Vora, A.R. Jani (2004) Superconducting state parameters of metals. *Indian J. Phys.* **78** 8 775-780.

[14] P.B. Allen, R.C. Dynes (1975) Transition temperature of strong-coupled superconductors reanalyzed. *Phys. Rev. B* **12** 905 DOI: https://doi.org/10.1103/PhysRevB.12.905.

[15] H.V. Khan, V. Singh, K.S. Sharma (1993) Application of pseudopotential theory for the prediction of superconducting state parameters of binary alloys. *Indian J. Pure & Appl. Phys.* **31** 9 628-634

[16] V. Singh, H. Khan, K.S. Sharma (1994) Screening dependence of superconducting state parameters of binary alloys. *Indian J. Pure & Appl. Phys.* **32** 12 915-924

[17] H.P. Dave, R.C. Malan, A.M. Vora (2018) Screening dependence of superconducting state parameters of binary alloys. *Acta Physica Polonica, A* **133** 12 86 DOI: http://doi.org/10.12693/APhysPolA.133.86.

[18] G. Sharma, S. Smita (2018) The pseudopotential dependence of superconducting properties in carbon doped MgB2. *AIP Conf. Proc.* **1953** 1 120044 DOI: https://doi.org/10.1063/1.5033109.

[19] K. Jasiewicz, B. Wiendlocha, P. Korbe, S. Kaprzyk, J. Tobola, (2016) Superconductivity of $Ta_{34}Nb_{33}Hf_8Zr_{14}Ti_{11}$ high entropy alloy from first principles calculations. *Phys. Stat. Sol. (RRL)?Rap. Res. Lett.* **10** 5 415-419 DOI: https://doi.org/10.1002/pssr.201600056.

[20] A.M. Vora (2006) Study of superconducting state parameters of alkali–alkali binary alloys by a pseudopotential. *Physica C* **450** 1-2 135-145 DOI: https://doi.org/10.1016/j.physc.2006.09.008.

[21] A.M. Vora (2007) Superconducting state parameters of Cu_xZr_{100-x} binary metallic glasses. *Physica C* **458** 1-2 43-50 DOI: https://doi.org/10.1016/j.physc.2007.03.397.

[22] C. Fiolhais, J.P. Perdew, S.Q. Armster, J.M. MacLaren, M. Brajczewska (1995) Dominant density parameters and local pseudopotentials for simple metals. *Phys. Rev. B* **51** 20 14001-14011 DOI: https://doi.org/10.1103/PhysRevB.51.14001.

[23] S. Korkmaz, S. D. Korkmaz (2006) A comparative study of electrical resistivity of liquid alkali metals. *Comp. Mater. Sci.* **37** 4 618-623 DOI: https://doi.org/10.1016/j.commatsci.2006.01.001.

[24] Y. Waseda (1980) "*The structure of non-crystalline materials: liquids and amorphous solids.*" McGraw-Hill, New York

[25] J. Hubbard (1958) The description of collective motions in terms of many-body perturbation theory. II. The correlation energy of a free-electron gas. *J. Proc. R. Soc. Lond. A* **243** 1234 336-352 DOI: https://doi.org/10.1098/rspa.1958.0003.

[26] L.J. Sham (1965) A calculation of the phonon frequencies in sodium. *J. Proc. R. Soc. Lond. A* **282** 1392 33-49 DOI: https://doi.org/10.1098/rspa.1965.0005.

[27] P. Vashishta, K.S. Singwi, (1972) Electron correlations at metallic densities. V. *Phys. Rev. B* **6** 3 875 DOI: https://doi.org/10.1103/PhysRevB.6.875.

[28] R. Taylor (1978) A simple, useful analytical form of the static electron gas dielectric function. *J. Phys. F Metal Phys.* **8** 3 1699 DOI: https://doi.org/10.1088/0305-4608/8/8/011.

[29] A. Sarkar, D. Sen, S. Haldar, D. Roy (1998) Static local field factor for dielectric screening function of electron gas at metallic and lower densities. *Mod. Phys. Lett. B* **12** 16 639-648 DOI: https://doi.org/10.1142/S0217984998000755.

[30] S. Ichimaru, K. Utsumi (1981) Analytic expression for the dielectric screening function of strongly coupled electron liquids at metallic and lower densities. *Phys. Rev. B* **24** 12 7385 DOI: https://doi.org/10.1103/PhysRevB.24.7385.

[31] B. Farid, V. Heine, G. Engel, I. Robertson, (1993) Extremal properties of the Harris-Foulkes functional and an improved screening calculation for the electron gas. *Phys. Rev. B* **48** 16 11602 DOI: https://doi.org/10.1103/PhysRevB.48.11602.

[32] I. Nagy (1986) Analytic expression for the static local field correlation function. *J. Phys. C Solid State Phys.* **19** 22 L481 DOI: https://doi.org/10.1088/0022-3719/19/22/002.

[33] H. Frohlic (1950) Theory of the Superconducting State. I. The Ground State at the Absolute Zero of Temperature. *Phy. Rev.* **79** 5 845 DOI: https://doi.org/10.1103/PhysRev.79.845.

[34] R. Sharma, K. S. Sharma (1983) Pseudopotential dependence of superconducting state parameters of Al. *Indian J. Pure App. Phys.* **21** 12 725-726

[35] R. Sharma, K. S. Sharma and L. Dass (1986) Pseudopotential dependence of superconducting state parameters of Al. *Indian J. Pure App. Phys.* **A60** 5 373-380

[36] S. C. Jain, C M Kachhava (1980) Superconductivity in certain metals *Indian J. Pure App. Phys.* **18** 7 489-493

[37] S. C. Jain, C M Kachhava (1981) Superconductivity in certain metals *Indian J. Pure App. Phys.* **A55** 89

[38] P.B. Allen (1999) *"Handbook of Superconductivity (ed) C P Poole (Jr.)"*. Academic Press, New York.

[39] B.T. Matthias (1951) "*Progress in Low Temperature Physics, 2, (ed) C.J. Gorter*". North Holland Publishing Company, Amsterdam.

[40] B.T. Matthias (1973) Criteria for superconducting transition temperatures *Physica.* **69** 1 54-56 DOI:
https://doi.org/10.1016/0031-8914(73)90199-7.

Index

A

acceptor-level, 52
alkali atoms, v, vii, viii, ix, 1, 2, 3, 5, 7, 11, 14, 15, 16, 18, 26, 27, 29, 30, 31, 32, 59, 60, 61, 73, 74, 77, 78, 80, 81
alkali metal, v, vii, viii, ix, 1, 3, 14, 21, 24, 25, 26, 33, 34, 35, 45, 47, 49, 50, 51, 52, 53, 54, 70, 83, 84, 85, 87, 88, 89, 91, 92, 93, 94, 95
alkali metal atoms, vii, 1, 21, 24, 25, 70

B

band structures, 52

C

$CH_3NH_3PbI_3$, viii, 33, 34, 35, 42, 45, 49, 51, 54
co-addition, 51
collision, vii, ix, 1, 3, 11, 14, 17, 20, 28, 31, 59, 61, 62, 64, 65, 67, 68, 69, 70, 73, 75, 77, 79
complex cross sections, 10, 11, 14, 17, 22, 29, 62, 65, 67, 76, 78
conduction band minimum, 53
conversion efficiency, viii, 33, 37, 51, 54
copper (Cu), 34
cross section, ix, 2, 3, 4, 7, 8, 9, 14, 17, 18, 20, 21, 24, 25, 26, 27, 29, 30, 31, 32, 60, 61, 62, 63, 64, 65, 67, 68, 69, 70, 71, 73, 75, 76, 77, 80, 81
crystal orientation, 39, 53
crystallite size, 47, 53
crystallization, 53
CsI, 46, 47, 49, 51
$CuBr_2$, viii, 34, 46, 47, 49, 50, 51, 53, 54
current density–voltage (J-V) measurements, 36

D

defect, 42, 43
density functional theory, 42
density of states (DOS), 42
diffusion coefficient, 5, 45
d-orbital, viii, 34, 52, 54

E

EA, 34, 51, 52, 53
electronic structures, 35, 42, 45, 49, 50
energy dispersive X-ray spectroscopy (EDX), 36, 41
energy gap (E_g), 38, 39, 42, 47, 49, 50, 51, 53
enthalpy, 44, 50
entropy, 50, 94
external quantum efficiency (EQE), 36, 39, 42, 46, 47, 49

F

Fermi level, 49
fill factor (FF), 37, 38, 41, 46, 47, 51, 52
first-principles calculations, 34, 51
formamidinium, viii, 33, 34, 35
frequency shifts, 2, 3, 5, 6, 8, 9, 10, 11, 13, 15, 16, 21, 24, 25, 27, 29, 30, 31, 80

G

GA, 34, 52, 53
Gibbs energies, viii, 34, 42, 54
growth mechanism, 44

Index

H

highest occupied molecular orbital (HOMO), 42, 49, 50
hole transport layer (HTL), 37

I

indirect optical orientation, ix, 59, 61
interaction potentials, ix, 2, 8, 9, 11, 14, 21, 26, 29, 60, 65, 66, 68, 69, 70, 72, 73, 74, 75
interstitial sites, viii, 33, 54

J

Jahn–Teller effect, 50

K

KBr, 35, 37, 38, 39, 40, 41, 42, 45
KCl, 35, 37, 38, 39, 40, 41, 42, 45
K-Rb System, 11

L

lattice constant, 38, 39, 41, 47, 50, 53, 54
lattice structure, 41
lowest unoccupied molecular orbital (LUMO), 42, 49, 50

M

magnetic measurements, vii, 1
magnetic resonance, v, vii, ix, 1, 2, 3, 8, 9, 10, 11, 12, 13, 14, 15, 16, 17, 18, 19, 20, 21, 22, 23, 24, 25, 26, 27, 28, 29, 30, 31, 32, 59, 60, 61, 62, 69, 70, 73, 74, 75, 76, 77, 79, 80, 81
magnetic resonance frequency, vii, ix, 1, 2, 3, 9, 10, 11, 12, 13, 14, 15, 16, 17, 18, 19, 20, 21, 22, 23, 24, 26, 27, 28, 29, 30, 31, 32, 60, 62, 69, 70, 74, 76, 80, 81
magnetic resonance frequency shifts, viii, ix, 2, 3, 10, 13, 15, 16, 19, 21, 22, 24, 26, 27, 28, 29, 32, 60, 74, 81
magnetoencephalographs, vii, viii, 1, 59, 61

N

Na, viii, ix, 14, 15, 16, 17, 30, 31, 33, 49, 50, 51, 52, 53, 54, 80, 83, 84, 87, 89, 91, 92
NaI, 46, 47, 49

O

open-circuit voltage (V_{OC}), 37, 38, 46, 47, 49, 51, 52, 54
optical microscopy (OM), 36
optical orientation of atoms, vii, viii, 1, 3, 18, 25, 26, 31, 59, 61, 64, 70, 79
optical pumping, vii, 1, 9, 10, 12, 13, 19, 20, 21, 22, 23, 28, 29, 30, 78

P

Perovskite, v, viii, 33, 34, 35, 50, 54
potassium, 8, 13, 14, 17, 20, 29, 30, 31, 34, 41, 78, 80
pseudopotential theory, 84, 85

Q

quantum gyroscopes, vii, viii, 1, 18, 59, 61
quantum magnetometers, vii, viii, 1, 2, 18, 59, 60, 61

R

RbI, viii, 34, 46, 47, 49, 51, 54
reaction rate, 45